THE HISTORY OF SCIENCE
IN THE NINETEENTH
CENTURY

ON THE SHOULDERS OF GIANTS

THE HISTORY OF SCIENCE IN THE NINETEENTH CENTURY

Ray Spangenburg and Diane K. Moser

Facts On File, Inc.

On the cover: Michael Faraday in his laboratory (AIP Emilio Segrè Visual Archives)

The History of Science in the Nineteenth Century

Copyright © 1994 by Ray Spangenburg and Diane K. Moser

Facts On File, Inc.
132 West 31st Street
New York, NY 10001

Spangenburg, Ray, 1939–
   The history of science in the 19th century / by Ray Spangenburg
and Diane K. Moser.
     p. c.m. — (On the shoulders of giants)
   Summary: Examines the role of science in the Industrial
Revolution, its establishment as a popular discipline, and
discoveries in the areas of atoms and the elements, chemistry,
evolution, and energy.
   ISBN 0-8160-2741-2 (alk. paper)
   1. Science—History—19th century—Juvenile literature.
   [1. Science—History.] I. Moser, Diane, 1944– . II. Title.
III. Series: Spangenburg, Ray, 1939– On the shoulders of giants.
Q125.S737 1994
509′.034—dc20   93-10576

Text design by Ron Monteleone
Cover design by Semadar Megged
Composition by Facts On File, Inc./Robert Yaffe
Manufactured by the Maple-Vail Book Manufacturing Group
Printed in the United States of America

MP FOF 10 9 8 7 6 5

This book is printed on acid-free paper.

In Memory of
Charles Darwin
and
T. H. Huxley

These two were very great men. They thought boldly, carefully and
simply, they spoke and wrote fearlessly and plainly, they lived modestly and
decently; they were mighty intellectual liberators.

—H. G. Wells in 1934

C O N T E N T S

Acknowledgments viii

PROLOGUE ix

INTRODUCTION THE GREAT AGE OF SYNTHESIS:
THE NINETEENTH CENTURY xi

PART ONE THE PHYSICAL SCIENCES 1
1. Atoms and Elements 3
2. Chemistry's Rollercoaster of Complexity and Order 21
3. Indestructible Energy 35
4. Magnetism, Electricity and Light 45
5. Sky and Earth 59

PART TWO THE LIFE SCIENCES 73
6. Darwin and the *Beagle*'s Bounty 75
7. From Macro to Micro: Organs, Germs and Cells 91

Epilogue 105
Appendix 1: The Scientific Method 107
Appendix 2: Elements and Their Symbols 111
Chronology 113
Glossary 120
Further Reading 123
Index 131

ACKNOWLEDGMENTS

*B*ooks are a team effort, and this one is no exception. Many people have been generous with their time, talents and expertise in helping us with this book. We'd like to thank everyone for their help and encouragement, and a special thank you to: Gregg Proctor and the rest of the staff at the branches of the Sacramento Public Library for their tireless help in locating research materials. Dorothea Nelhybel at the Burndy Library for her willingness to help with photos and illustrations despite the fact that her library was in the middle of moving. Beth Etgen, educational director at the Sacramento Science Center, and her staff, as well as science historian Karl Hall, for kindly reading the manuscript and making many helpful suggestions. Andrew Fraknoi of the Astronomical Society of the Pacific, for his help with research contacts and illustrations. Also for their help with illustrations: Leslie Overstreet of the Smithsonian Institution Libraries, Diane Vogt-O'Connor of the Smithsonian Archives, Jan Lazarus of the National Library of Medicine, Clark Evans of the Library of Congress, R. W. Errickson of Parke-Davis and Doug Egan of the Emilio Segrè Visual Archives. Thanks above all to Facts On File's editorial team, especially Nicole Bowen for her intelligent, energetic and up-beat management of the project; Janet McDonald for her eagle-eyed passion for detail; and James Warren for his vision and encouragement in getting it started. And to many others, including Jeanne Sheldon-Parsons, Laurie Wise, Chris McKay of NASA Ames, Robert Sheaffer and Bob Steiner, for many long conversations about science, its history and its purpose.

PROLOGUE

*I*t's sometimes called the Golden Age of Science, an era when science seemed to be at the forefront of human activity and scientists were making great and exciting advances, both applied and theoretical. The 19th century was a time of major breakthroughs and leaps toward a fuller understanding of nature in both the physical and life sciences; a time when science came into its own and commanded the attention of society, giving impetus to technology, new perspectives to sociology, and catalytic stimulation to the arts. But what did 19th-century men and women of science mean when they used the term *science?*

In the 17th and 18th centuries, many scientists became adamant about a method of solving problems that came to be known as the scientific method. Many argued that it was the only right way to arrive at sound solutions:

1. Observe

2. Develop a hypothesis from which predictions can be made, which, in turn, can be tested

3. Develop and do experiments that can offer irrefutable proof.

(For more about the methods of science, see Appendix One at the back of this book.)

But by the beginning of the 19th century, scientists had begun to find that there was more than one fruitful approach to a problem. Some problems responded better to one type of attack, some to another.

In England, especially, theorists became fond of using a model to represent how one might think of a structure or relationship. The model wasn't intended to represent reality literally. It just helped visualize the interaction of molecules in a chemical reaction or the steps of a physiological process.

Intuition also paid off sometimes. Many scientists begin by working on a hunch. Or they may begin a particular line of reasoning for reasons as vague as it "seems right."

By the end of the century, even mathematical proofs sometimes won the day, for lack of any more tangible experimental method for tackling a problem.

But the big thing to remember, regardless of approach, is that science is self-correcting. That's what makes science different from other human searches for truth. It's a lot like working on a giant jigsaw puzzle. If, in the early stages of the puzzle, you put a piece in the wrong place, you'll eventually see where you've gone wrong as the rest of the picture emerges. You can spot where you've placed a piece of shrub in the middle of the sky, or attached a horse's tail to a pollywog. And you can change the mispositioned pieces around.

But the out-of-place puzzle piece can also be productive. Other pieces that fit with it may cluster around it and the whole assemblage may be moved to a new spot in the picture that makes it all work. And this is one of the fascinating aspects of the dynamic of science as it grows—one we have had to leave out of this book for the most part, unfortunately, for lack of space. It's easy to get the impression in reading a history of science that everything moves surely forward, one discovery leading smoothly to another. In fact, science moves by fits and starts, with volumes of data collected that may seem contradictory or appear to lead nowhere. Many perfectly sensible-sounding theories get offered that are cast aside in favor of another—sometimes rightly, sometimes wrongly. The cast-offs are sometimes resurrected later. Or some piece of a "wrong" theory may suddenly be seen to fit with a piece of another. It all contributes to the pot of knowledge and it's all part of the dynamics of an expanding understanding of the universe and how it works.

Although we have time here only to follow the main threads, for those who would like to know more about these adventures in discovery and insight, about rich, productive lives and what they have produced, we'd like to recommend turning to some of the books listed at the back of this book. And we wish you many hours of exciting exploration!

The History of Science in the Nineteenth Century is one of a series of five books, called *On the Shoulders of Giants*, which looks at how people have developed the methods of science as a system for finding out how the world works. Each book in the series looks at the theories they put forth, sometimes right and sometimes wrong, and at how we have learned to test, accept and build upon those theories—or to correct, expand or simplify them.

Each book also explores how scientists have learned from others' mistakes, sometimes having to discard theories that once seemed logical but later proved to be incorrect, misleading or unfruitful. In all these ways these men and women—and the rest of us as well—have built upon the shoulders of the men and women of science, the giants, who went before them.

INTRODUCTION

THE GREAT AGE OF SYNTHESIS: THE NINETEENTH CENTURY

We look for a bright day of which we already behold the dawn.
—Humphry Davy, in a lecture in London

But though this consummation may never be reached by man, the progress of science may be, I believe will be, step by step towards it, on many different roads converging towards it from all sides.
—William Thomson (Lord Kelvin), physicist

A rmed with the continued success of science and its applications, scientists greeted the 19th century with great optimism. The year 1800 marks the dawn of a time when expanding scientific knowledge and technological advancement virtually created the positive, can-do spirit of the Victorian era that spanned the last two-thirds of the century. By the end of the century, gaslights lit the streets of London, telegraph communication transformed journalism and business alike, factories hummed and urban streets bustled with commerce.

Politically and economically, however, the sense of the time was not so monolithically positive. The 19th century was a time of alternating periods of peace and revolution, of nationalism in Europe, in the Ottoman Empire and in the Americas, of internal struggles to liberalize the governments of Europe, of industrialization and accompanying growing pains, and of wide-ranging imperialism on the part of the countries of Europe.

The first years of the century were plagued by the expansionism and wars of France's emperor, Napoleon Bonaparte. In 1815, with Napoleon defeated at last, the governments of Europe met at the Congress of Vienna to re-establish peace and a balance of power to Europe. But revolutions broke out between 1820 and 1830. The year 1848 brought widespread revolution—partly as a result of the economic crises of 1846–48, caused by the great potato famine in Ireland, by a small grain harvest due to drought, across Europe, and by the economic hard times that followed. Food shortages, high prices and unemployment all set the stage for revolt.

In addition, many leaders had backtracked on the civil rights symbolized by the French Revolution. People were angry and hungry and tired of being powerless, and socialist ideals gained wide appeal among the workers of France. The Communist Manifesto written by Karl Marx and Friedrich Engels, first published in early 1848 and translated into French, added further fuel to the growing unrest. In a bloody French workers' revolt in June of that year, at least 1,500 people were killed in the uprising, as many as 8,500 were wounded and thousands wound up in jail. But the right to vote was won for every male—not just landowners—in what constituted a major milestone in the struggle for equal rights (though voting rights for women still lay a long way off). Other revolts broke out in Austria, Hungary and Italy, as well.

Meanwhile, under the influence of industrialization and the voracious hunger it produced for cheap raw materials, worldwide imperialism became stronger. Europeans took over nearly every country in Africa by the end of the century, and Britain established sovereignty in India and a vast array of far-flung lands. And the Far East, including China and Japan, were opened up to trade (not always without bloodshed).

Then, as now, science sometimes served governments in time of war, but for the most part the pursuit of science stood apart from international politics and yet at the heart of society's economic growth and industrialization. The 19th century saw the rise of a new kind of alliance, an international intellectual alliance of seekers of truth, that cut across national boundaries and overcame the parochial concerns of nation against nation.

It was a time, too, when professionalism came to science—when more and more men and women made their living by their scientific work—and when the fruits of science (as well as the hazards) played an increasingly prominent role in the public mind. Geologists transformed mining. Physicists brought new insights into ways to harness energy. And advances in the biological sciences made possible important breakthroughs in medicine and health science.

The Industrial Revolution, the most wide-ranging effect of scientific discovery, completely transformed the way people lived and worked in the

19th century. Begun in the 18th century, especially in England, with inventions that mechanized the production of cloth, the Industrial Revolution moved into full swing in the 19th as James Watt's steam engine (perfected in 1781) found more and more uses in industry and transportation. By 1804, Richard Trevithick, in England, had built a locomotive that pulled five loaded coaches along a track for nine and a half miles, and by 1814, the great railroad pioneer George Stephenson had introduced his first steam locomotive. Soon railroads crisscrossed England, Europe and North America as factory production increased with the use of steam power and the demand for efficient transportation escalated. Urban centers took on more and more importance in what had once been bucolic countryside. Overall, throughout the century, science lay close to the heartbeat of all progress, a catalyst of industrial and intellectual change.

Industrialization also became a tool of greed, however, and took the blame for hardships suffered by workers whose employers required them to work long hours in unsafe conditions for too little pay. Not everything about the Industrial Revolution was glorious, and those who lost jobs to new inventions and endured inhumane conditions rapidly came to resent the growth of technology and science.

But industrialization overall boosted quality and availability of goods for an enormous number of individuals. The improvements in transportation increased mobility, connected isolated communities and improved the flow from farm to market and from factory to market. In England in particular, improved economic conditions brought new opportunities and a better quality of life, especially for the middle class.

Science, as a result, loomed large in the public mind, and from the mid-1830s on, lectures on science became enormously popular. In late November 1859, a book by biologist Charles Darwin called *On the Origin of Species* sold out on the first day of its publication. Associations and societies for those interested or engaged in science sprang up everywhere. An organization known as the British Association for the Advancement of Science was founded in 1831 and a man named William Whewell coined the word *scientist* in 1840, replacing *natural philosopher*, to describe its members. The American Association for the Advancement of Science was founded in 1848, and journals devoted to science, though mostly applied science, gained great popularity in the United States.

This was the era in which science and scientists came of age. Experimental approaches and procedures became much more complicated (a trend that has continued in the 20th century), to the point that by the end of the century the era of the amateur scientist had ended. For the first time, out of necessity, scientists were primarily full-time professionals, usually specialists, not part-time amateurs or generalists. They began to need outside financial support even just to obtain the equipment for their experiments. And they required formal training to keep up with their fields, which became more and more specialized into specific disciplines such as chemistry, physics, astronomy, biology and subdisciplines such as organic chemistry and genetics.

Scientists also began to specialize between theorists and experimentalists, especially in the physical sciences. This was not an entirely new trend. Johannes Kepler, the great astronomer-theorist of the 17th century, had stood on the shoulders of Tycho Brahe, the great astronomical observationist, who had collected the mountains of data from which Kepler drew his conclusions. Isaac Newton, the great synthesizer, stood on the shoulders of Galileo, the experimenter. But, with the 19th century, the interplay of experiments and theories became even more pronounced. And the roles of experimenter and theorist soon rarely resided in the same person—there was simply too much to be done, too broad a swath to cut. And the approaches required became vastly different—too different, as a rule, to reside comfortably in the same personality. How many people could possess both the fastidious attention to detail and tenacious perseverance required of a good experimenter and, at the same time, the theorist's ability to think broadly and abstractly, juxtaposing seemingly unrelated concepts, and interpreting and synthesizing results?

It was also a time of increasing complexity in the substance of science—not just in specific disciplines such as chemistry, physics, astronomy, biology, psychology, organic chemistry. Both the fields of chemistry and geology reached a new maturity after the striking advances of the previous century. The boundaries between the sciences became more or less set, and the almost "Renaissance," multi-disciplinary approaches of 18th-century investigators like Joseph Priestley, René Descartes and Benjamin Franklin gave way to specialization. In the 1830s, John Herschel could still choose to be a generalist, making contributions not only in astronomy, but in chemistry and mathematics as well. But he was already an exception to the rule. Science had become too complex for an individual to make significant contributions without going deep into a single area or discipline.

Yet the 1800s were also a great time of synthesis in the sciences. From the time of the Greeks, scientists have looked for a few simple, underlying principles to explain the seemingly unrelated, complex details of the physical universe and the living organisms that inhabit it. By the 19th century the idea of convergence had become paramount. Strong indications had begun to emerge that everything might be explained by just a few explanatory theories—if indeed not by just one.

Physicists, especially, had already had more than a taste of evidence that everything might soon converge into one. Isaac Newton had shown in the 17th century that the same force caused the fall of an apple in an orchard and the periodicity of the Moon, two events that seemed at first glance (and had for many centuries) not to be remotely related. Benjamin Franklin, in the 18th century, had shown that the static shock one might get from an iron railing was related to lightning in the thunderclouds overhead.

"I wish we could derive the rest of the phenomena of nature by the same level of reasoning from mechanical principles," wrote Newton, "for I am inclined by many reasons to suspect that they may all depend on certain forces."

The scientists of the 19th century were primed to find even more convergence than Newton probably meant to suggest, to come up with even more all-encompassing ideas. And they did.

Atomism, which came of age in the early 1800s under the direction of John Dalton, was above all a reductionist idea—the desire to reduce all the complex forms of matter occurring in nature to a few fundamental particles that respond to a few basic laws.

In 1799 Alessandro Volta put together his "voltaic cell," the first usable battery. Before his invention, scientists could neither really study electricity nor use it because they could only get fleeting glimpses of small amounts of static electricity or momentary discharges. Now they had an electric current. Then Hans Christian Ørsted discovered by accident that electricity and magnetism were linked! He published his discovery in 1820. Major breakthroughs tumbled one after the other out of the laboratories and calculations of Michael Faraday, André Marie Ampère and others. In a striking example of how a new scientific tool can uncork a bottleneck, much of 19th-century science flowed from the electromagnetic theory that followed and from the use of electrolysis in chemistry.

The stunning idea that electricity, magnetism and light were all forms of the same energy force galvanized the world of physics. In fact, energy, investigators found, could be transformed into many different forms: heat, mechanical motion, electricity, light. Many scientists believed that energy would be the great unifying theme of the century—that the answer to everything, ultimately, would boil down to one unifying theory of energy. So much progress was made in the 1800s, in fact, that many physicists believed that only a few problems remained to be solved. The study of physics, they claimed, would soon come to an end—since not much was left to discover.

Great, all-encompassing ideas were not confined to the physical sciences, either. In biology, two men, Charles Darwin and Alfred Wallace, came up with an extraordinary set of principles explaining the origin of the species. The theory of evolution would provide viable explanations for the enormous diversity of life that became wider and more mind-boggling with each new voyage to far-away, isolated lands. And Gregor Mendel's laws of genetics offered new insights into how traits were passed on from one generation to another.

However, not everyone believed that unity in science would come through convergence of theories. Some, like James Clerk Maxwell, thought instead that the sciences would be unified by method of approach, not by any one theory (strangely enough, since he was the author of electromagnetic theory, one of the great unifying concepts of all time). In England, especially, the most frequently used method was the idea of using an analogy, or model, to represent how a concept worked. (The French tended to think that the approach was childish and simplistic, but in England, scientists of varied backgrounds found that a fountain of ideas flowed from building a mechanical model. Dalton, Faraday, William Thomson and Maxwell all found models enormously fruitful.)

The 19th century also saw the demise of alchemy and its mysticism, which had dogged much progress, especially in chemistry, for many preceding generations. No longer did chemists speak, by the end of the century, of a mysterious group of substances known to their predecessors as "imponderables." For the previous century, this vestige of alchemy had dominated almost every effort to get at the nature of reactions that took place when substances were combined or when combustion took place. Heat, light, magnetism and electricity were all thought to be weightless fluids that transferred from one substance to another. Their presence could not be detected by weight because they had no weight. Lavoisier had discredited the theory of "phlogiston," another imponderable postulated to explain combustion, but the rest of the imponderables remained part of scientific theory until, in the 19th century, one by one, discoveries led to sounder explanations of the way things worked. And the last traces of mysticism finally dropped away from scientific inquiry.

As scientific ideas became more powerful and cohesive, controversy swirled around them. Some criticized the rejection of long-held beliefs, including alchemy, mysticism and astrology. Many balked at new theories—Darwin and Wallace's theory of evolution, in particular—that seemed to compete with biblical accounts and overthrew the long-accepted hierarchy that set humans apart from the rest of the animal world.

Several influential factions of European society also opposed science outright as anticreative, rigid and constrictive. In Germany, the scientist-poet Johann Wolfgang von Goethe and the philosopher Georg Hegel became arch-opponents of what they termed the mechanistic and materialistic nature of science, and the idealistic and romantic German *Naturphilosophie* gained popularity toward the end of the 18th and beginning of the 19th century. In France, following the restoration of the Bourbons in 1814, antiscientific Romanticism became the socially popular stance to take. Such prestigious figures as the writers Anne Louise Germaine de Staël (often referred to as Mme de Staël [duh stah EL]) and René de Chateaubriand [shah TOH bree anh] scorned the "whole brood of mathematicians," and the French poet Alphonse de Lamartine, intoxicated over the power of human emotion, wrote loftily, "Mathematics were the chains of human thought. I breathe, and they are broken." Romantics saw human feeling and individualism as the source of all creative power, and they saw the universe as an organism, not a machine. They set the subjectivity of the "heart" and imagination in opposition to the more objective fruits of the scientific mind. The English poet John Keats spoke for many Romantics when he wrote that he was "certain of nothing but the holiness of the heart's affections, and the truth of imagination. What the imagination seizes as beauty must be truth . . ."

One of the greatest controversies of the era centered around evolutionary theory, which postulated that the diverse species in nature had evolved from common ancestors through natural selection. Nearly every conservative paper in England ran cartoons lampooning Darwin and his ally, T. H. Huxley, as apes, monkeys or gorillas. But the attention given issues like this in the press marked the high level of significance they had in the popular mind. Long past were the days of Copernicus, in the 16th century, when only a few educated scholars had any hope of following the arguments set forth by science, let alone any interest. These were exciting times, when science, or at least a public persona of science, held a center-stage spotlight for all to see. It was a spotlight whose intensity increased as the years—and discoveries—rolled forth.

But at the start of the 1800s, it all began with a few simple inventions and discoveries—a theory set forth by a solitary teacher named Dalton in England, and a battery invented by a lone physicist named Volta in Italy. With these two events, our story begins.

PART ONE

THE PHYSICAL SCIENCES

C H A P T E R 1

ATOMS AND ELEMENTS

Nothing exists except atoms and empty space;
everything else is opinion.
—Democritus

John Dalton's dark, broad-brimmed hat and somber clothes silhouetted starkly against the gray skies of northern England as he took his daily hikes through the hillsides around Manchester. As the smoke from the growing industrial city's stacks mixed with the dank fog of the moors, Dalton took meticulous meteorological measurements, noted the continual changes of the Lancashire County weather, made studies of the atmosphere and performed experiments. To his neighbors he seemed an odd and solitary figure. But to science this reclusive, scantily educated, individualistic Quaker became one of the great catalysts of the early 19th century, formulating one of the most basic theories of modern science, a theory that would become the foundation for modern chemistry and physics.

To John Dalton goes the credit for reestablishing the ancient idea of the atom, to which he first alluded at the end of a paper in 1803, and by 1826 his renown was widespread. In that year, at a meeting of the Royal Society, Britain's prestigious society of scientists, chemist-physicist Humphry Davy (whom we'll meet later in this chapter) proclaimed:

> *Mr. Dalton's permanent reputation will rest upon his having discovered a simple principle, universally applicable to the facts of chemistry . . . and thus laying the foundation for future labours. . . . His merits in this respect resemble those of Kepler in astronomy.*

John Dalton was not, however, the first to come up with the idea.

John Dalton, whose atomic theory opened the door for much of 19th-century science (AIP
Emilio Segrè Visual Archives)

NATURE'S BUILDING BLOCKS

For many centuries most ancient peoples speculated that all substances were
made of a few basic elements. The Greeks thought in terms of four
fundamental substances that they called elements: air, fire, water and earth.
The Hindus subscribing to Ayurvedic philosophy imported the four-ele-
ment theory from the Greeks, and ancient Chinese Taoists developed a
theory of five interacting elemental "phases": metal, water, wood, fire and
earth.

But most of these ancient philosophies did not include the idea of atoms. Sometime in the 5th century B.C., a lone Greek thinker named Leucippus [lyoo SIP us] wondered what would happen if you broke a substance down to its smallest possible particle. What, for instance, if you split a rock in half, then split that in half and again in half and again and again and again. Soon (sooner than you might suppose) you would have particles of dust. Could you split a bit of dust in half? Yes, speculated Leucippus (although, as far as we know, he did not try it). And then could you split it again in half? How far could you go? Eventually, Leucippus thought, you would reach the smallest possible particle, and this hypothetical mite, too small to be seen, he called an *atom*, the Greek word for unsplittable.

His student Democritus (c. 470–c.380 B.C.) took up atomism as well and expanded on Leucippus's theory, maintaining that nothing but empty space existed between atoms and that all things, including the human mind, were composed of atoms, which moved mechanistically, according to laws of nature. But familiar as these ideas sound today, Leucippus and Democritus arrived at their conclusions not by experiment, but as the Greeks usually did, by considered reasoning. The Arab scientist Rhazes (ar-Razi, born c. A.D. 852) later held atomic views similar to those of Democritus and maintained that atoms gave rise to the four elements. By the 11th century scientists in India had developed a unique atomic theory, with combinations formed in dyads (combinations of two) and triads (combinations of three).

In the 17th century Robert Hooke (1635–1703) thought that the pressure exerted by a gas against the sides of a container (a balloon, for example) might be caused by a rush of atoms milling around. His contemporary Robert Boyle (1627–91) recognized early that gases were probably the key to understanding atoms (which he liked to call "corpuscles"). He showed, in a famous experiment with a J-shaped tube, that air could be compressed. One good explanation for this, he thought, could be that the atoms in a gas were usually widely separated by empty space but moved closer together under pressure. It wasn't proof that atoms existed, though; a lot of other explanations were also possible.

Later, scientists in the 18th century discovered that water was composed of two elements, hydrogen and oxygen (so water wasn't an element after all). And they discovered what they referred to as different kinds of "air," gases that we now call oxygen, nitrogen and carbon dioxide. Other elements had also been discovered by that time, too, and so the ancient ideas about how many elements there were and their identity no longer seemed valid. But the basic idea that all substances are composed of a relatively small number of elements still held.

The idea of the atom, though, did not generally catch on among most scientists at first. For one thing, two influential Greek thinkers, Plato (c. 427–347 B.C.) and Aristotle (384–322 B.C.) did not subscribe to Leucippus and Democritus's idea, and though a few renegades existed, no one had ever shown compelling experimental evidence that pointed to the existence of atoms.

THE "NEW CHEMISTRY"

So the idea of the atom had been around a long time when John Dalton came on the scene, but no one had ever found a way to offer experimental proof that a tiny, invisible building block of matter existed. And they had also never found a way to explain the diverse chemical properties of the huge number of materials known even then.

Groundwork in chemistry had been laid in other ways, however. Antoine Lavoisier, Joseph Priestley and Joseph Black had shown that chemistry, like physics, could profit enormously from the use of measurement. Primarily by weighing before and after, they began quantifying the results of their experiments and showed that sound theories and conclusions could be built on the basis of quantitative analysis.

Lavoisier also developed what has become known as the law of conservation of matter—the idea that matter is neither created nor destroyed, but is simply transformed. And he proposed that chemical elements are no more, no less than substances that one cannot break down into any simpler substance by chemical means. By the end of the 18th century, moreover, chemists had discovered a host of new elements not recognized before.

But it would be John Dalton's atomic theory that would provide an explanation of the structure operating behind the scenes.

DALTON'S ATOMS

As a young man John Dalton (1766–1844) hardly seemed destined to shake up the scientific world. He was never a great experimentalist. He was neither brilliant nor eloquent, and he had no access to the "best" schools. As a child he attended a one-room school, and at the age of 12 he took over teaching the entire school. In his spare time he read books by Newton and Boyle, and from that point on, for the most part, he was self-taught, unable to afford extended schooling. Soon afterward Dalton opened up his own school, but he was such a boring lecturer that within three years the school closed because all his students had dropped out.

Unlike many scientists of his day, Dalton never had much success on the lecture circuit. He had a gruff, country way about him, and his blunt manner completely lacked charisma. So, unable to pick up income as a lecturer, he supported himself most of his life as a teacher and tutor, pursuing his scientific interests in every remaining moment. (When asked why he never married, he replied dryly, "I haven't time. My head is too full of triangles, chemical processes and electrical experiments to think of any such nonsense.")

A Friend, or Quaker, he strictly followed his religion's custom of plain dress, which may have worked to his advantage, since he was color blind, and the choices of somber clothing in his closet included none in colors he could not see.

John Dalton had a great persistence about him, a methodical, consistent approach and an unflagging curiosity. He kept a daily meteorological diary his entire life from the year 1787 on, made several contributions to the study of gases, and gave the first clear statement of atomic theory—all part of his relentless delving into the mysteries of nature.

Many unsolved questions remained at the end of the 18th century about the nature of air and its components, and Dalton was fascinated. He made some 200,000 meteorological observations during his daily excursions over the course of his life—the last on the day of his death at age 78. It's little wonder that his involvement in the study of the weather led him to explore the mysteries of gases, their behavior and composition.

Air is composed mostly of oxygen, nitrogen and water vapor, that much was already established. But why didn't the various parts of the mixture always separate out? Why didn't the heavier gas, nitrogen, sink to the bottom of a container, or, for that matter, to the lower regions of the atmosphere? Using simple, home-built apparatus, Dalton weighed the different elements of which air is composed and came to some important conclusions.

Dalton discovered that a mixture of gases weighed the same as the combined weight of the gases taken separately. As he explained it:

> When two elastic fluids [gases] denoted by A and B are mixed together, there is no mutual repulsion between their particles; the particles of A do not repel those of B as they do one another. Consequently the pressure or whole weight upon any one particle arises solely from those of its own kind.

Known as Dalton's law of partial pressures (announced in 1803), basically the statement boils down to the idea that different gases in a mixture do not affect each other, or that the total pressure of a mixture of gases is the sum of the pressures of each gas taken singly. Dalton, of course, knew about Boyle's work with gases, and this new piece of information seemed to point even more strongly toward the idea that gases were made up of tiny, indivisible particles.

But he kept thinking about it. What if all matter—not just gases—was made up of these tiny particles? Joseph Louis Proust had pointed out in 1788 that substances always combine in whole units. That is, chemicals might combine by a ratio of four to three. Or eight to one. But the reaction would not take, say, 8.673 grams of oxygen and 1.17 grams of hydrogen. One way to explain this law of definite proportions, as it was called, was to assume that each element was composed of small, unsplittable particles.

The evidence was building up. So Dalton set forth the idea that each element was made of tiny particles, which, in honor of Democritus, he named "atoms." (This was a slightly confusing name, though Dalton couldn't know it, because, as we now know, atoms are not really "unsplittable." They are, in turn, made up of several even tinier particles that appear to be unsplittable. For this reason, many scientists today like to refer to Dalton's atom as the "chemical atom.") Dalton also proposed that atoms of different chemical substances were not the same, as some earlier atomists had claimed. But, unlike Democritus, who thought that the atoms of different substances differed in shape, Dalton observed that they differed in weight, and he established the fact that each element has a weight of its own, particular to that element.

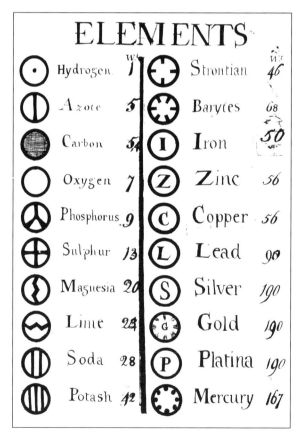

Dalton's original list of elements and symbols
(The Bettmann Archive)

AVOGADRO'S HYPOTHESIS

All gases, as Joseph Gay-Lussac showed definitively in 1802, expand to the same extent with a given increase in temperature. (John Dalton had also come to the same conclusion, and, independently, a man named Jacques Charles had anticipated both of them. This principle of the constant rate of expansion of gases at a constant pressure is now known as Charles's law, since Charles came up with it first.)

This law must mean, announced a man named Amedeo Avogadro (1776–1856, count of Quaregna) in 1811, that equal volumes of different gases (at the same temperature) must contain the same number of particles (notice he did not say "atoms"). This idea, called Avogadro's hypothesis, stirred up considerable controversy throughout the first half of the 19th century.

If this were so, why did a given volume of oxygen or hydrogen weigh twice as much as one would expect from the atomic weight of these gases? (Oxygen and hydrogen atoms we now know, occur in nature bound together in molecules of two—but in Avogadro's time this was not known.) If you chemically combine a volume of hydrogen and a volume of chlorine, Avogadro thought, you might expect to get one volume of hydrogen chloride gas. But instead you get two volumes. Did that mean that the hydrogen atoms and chlorine atoms were splitting to combine with each other? No, said Avogadro. Some elements, he hypothesized, might be combinations of atoms, and in fact, he thought that some gases—including oxygen, nitrogen and hydrogen—occurred naturally in molecules composed of two atoms (O_2, N_2, H_2). (Avogadro was the first to use the word *"molecule,"* which means "little mass," in this sense, and he was the first to distinguish in this way between atoms and molecules.) However, many major figures in chemistry—among them Dalton and the renowned Swedish chemist Jöns Jacob Berzelius—rejected the idea on the assumption that like atoms repel, and Avogadro's hypothesis remained dormant for many years, until 1858, when it finally became accepted.

In September 1803, Dalton presented his first list of atomic weights based on hydrogen as a unit of one, with all other elements weighing multiples of that weight. He later expanded his original list to include 21 elements.

Thanks to Dalton, chemists began to realize that there were different types of atoms and that atoms of any one element were all alike, with specific properties and differing in relative weight from the atoms of all other elements.

Dalton kept pushing on the idea. Two elements, it seemed, might combine to form more than one compound. Carbon and oxygen, for example, combine to form what we now call carbon monoxide and also carbon dioxide. But they combine in different proportions, still always whole numbers (a ratio of 3:4 by weight of carbon to oxygen in carbon monoxide and a ratio of 3:8 by weight in carbon dioxide). Dalton surmised that carbon monoxide might be just one particle of carbon combined with one of oxygen (with four carbon particles equal in weight to three of oxygen). Carbon dioxide, he figured, was one particle of carbon united with two of oxygen. (A hypothesis that was later confirmed.) Known as the law of multiple proportions, which Dalton published in 1804, this idea had been anticipated by a scientist named William Higgins in 1789, but no one had supported it with experimental evidence until Dalton came along. Many of Dalton's colleagues found it exciting because it seemed to make the atomic theory even more plausible.

Dalton came up with this idea because he noticed that when elements combined chemically to form a compound, one or more atoms of one element combined with one or a small number of atoms from the other to form a molecule, the smallest particle of a compound. For instance, a molecule of water is always composed of one part, by weight, of oxygen to two parts hydrogen. A molecule of water always has the same molecular weight as every other molecule of water, and it is the same as the weight of two atoms of hydrogen and one atom of oxygen. Dalton tested this out with several dozen compounds and always got the same results.

Dalton's theory of the atom made it possible to explain how these elements combined to form compound substances. Atoms got together, he said, to form other substances, and when they did, they combined chemically one with one, or one with two or three or so—always whole numbers, not split—to form other substances.

In 1808 he published his ideas in his *New System of Chemical Philosophy*. The atom, he declared, was the basic unit of the chemical element, and each chemical atom had its own particular weight. He wrote:

> *There are three distinctions in the kinds of bodies, or three states, which have more specially claimed the attention of philosophical chemists; namely, those which are marked by the terms elastic fluids [gases], liquids, and solids. A very famous instance is exhibited to us in water, of a body, which, in certain circumstances, is capable of assuming all three states. In steam we recognize a perfectly elastic fluid, in water a perfect liquid, and in ice a complete solid. These observations have tacitly led to the conclusion which seems universally adopted, that all bodies of sensible magnitude, whether liquid or solid, are constituted of a vast number of extremely small particles, or atoms of matter bound together by a force of attraction, which is more or less powerful according to circumstances. . . .*

He went on to explain that chemical analysis and synthesis simply involved reorganizing these particles—separating them from each other or joining them together. As Lavoisier had said, matter was neither created nor destroyed in the process. "All the changes we can produce," Dalton declared, "consist in separating particles that are in a state of cohesion or combination, and joining those that were previously at a distance." Insights, all, that still hold firm today.

Dalton's atomic theory found success where others had failed because he provided a model from which definite predictions could be made. Some features of his theory, certainly, were later set aside, but the core features survived: that each atom has a characteristic mass and that atoms of the elements remained unchanged by chemical processes.

Along the way Dalton came up with some other, less major discoveries. He was the first to publish the generalization that when any gas starts out at the same temperature as another gas, both will expand equally when heated to the same higher temperature. He was also the first to describe color blindness, in a paper published in 1794.

In 1833 a group of admirers and friends collected contributions to build a statue of Dalton, which was erected in front of the Manchester Royal Institution. Several prestigious societies honored him, including the Royal Society in London and the Academy of Sciences in Paris. And in 1832, when he received a doctorate from Oxford, he had the opportunity to be presented to the king of England. The only problem was he would be required to wear court dress, including a sword, which was directly contrary to the pacifist principles of his religion. But he and British dignitaries settled on a compromise: He would wear the cloak of Oxford, which would make the question of the sword a moot point. He may or may not have known that the cloak he wore was bright red—also completely out of keeping with Quaker custom. But to the color-blind scientist, the cloak was gray.

John Dalton died in 1844, much respected for his work on atomic theory and the behavior of gases. More than 40,000 people filed by the coffin of the man who couldn't seem to get students to come to his classes. He had set the stage early in his life for the greatest 19th-century developments in both physics and chemistry, and he was lucky enough to have seen the day when people recognized the value of his contribution.

THE ELECTRIC CONNECTION

The story will return to the saga of atoms and elements in chemistry, but first, a side trip to Luigi Galvani's laboratory in Bologna, Italy. It's the summer of 1771, a few years before our 19th-century story begins. The laboratory is a mess, with dozens of pairs of frog legs (possibly destined for a pot of soup, according to some accounts) strewn across the wooden table top.

VOLTA AND THE BIRTH OF THE BATTERY

Alessandro Volta began the age of electricity in 1800 when he announced his invention of the electric battery, the first continuous source of electric current. Not that studies of electricity hadn't been made before; scientists worldwide (including Benjamin Franklin) had made studies of static electricity for a century. But static electricity discharged all at one time, in a single spark or jolt. Volta's battery could provide a current, and while practical uses were not found for it for several years, chemists and physicists immediately set it to work as a tool in analyzing substances.

Volta, a physicist at the University of Pisa, got his first inspiration from the work of Galvani on frogs (see text), and did some experiments of his own. Skeptical about "animal electricity," he was struck by the fact that electricity was produced only if two different metals were used. Also, he noticed, some combinations of metals produced more twitching than others. And, when he placed tinfoil and silver on his tongue, which is mostly muscle, his tongue didn't twitch much, but he did notice a sour taste. This made him wonder if the electricity might be conducted from one metal to the other by the fluid of his saliva. Volta experimented with many solutions, finally settling on brine, which is a strong solution of saltwater. He found that if he constructed piles of dissimilar metal disks, sandwiching pieces of cardboard-soaked brine in between them, he had a very effective set of battery cells (each sandwich forming one cell). The resultant stack became known as the voltaic cell, named after its inventor.

Its potential was immediately recognized by scientists everywhere and even by Napoleon in France, who made him a count and a senator in Volta's native kingdom of Lombardy (which recently had been conquered by Napoleon). In a letter to the Royal Society in London in 1800, Volta expressed his own delight in his invention's wonderful simplicity. "Yes!" he wrote in his letter, "the apparatus of which I speak, and which doubtless will astonish you, is only an assemblage of a number of good conductors of different sorts arranged in a certain way."

And Alessandro Volta's astonishing "assemblage," often referred to as a "voltaic pile," became a tool as important to science as the telescope and the microscope long before it came to transform the way people lit their homes and streets.

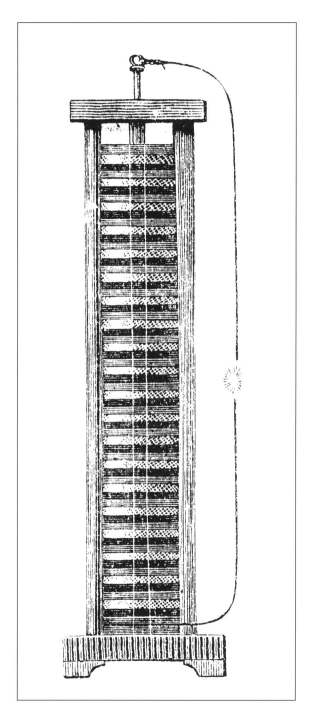

Alessandro Volta's "voltaic cell," the first source of continuous electrical current. It was an electrochemical cell that operated on much the same principle as today's flashlight batteries.
(The Bettmann Archive)

Luigi Galvani [gahl VAH nee] (1737–98) was an anatomist and physician, not a physicist. He was professor of biology at the University of Bologna. But it came to Galvani's mind to try stimulating the muscles of the dissected frog legs with a spark from an electric machine. The frog legs twitched on contact. He even found that a metal scalpel caused the legs to twitch if the machine was turned on—even if the spark didn't actually make contact.

If an electrical spark caused this muscle twitching, Galvani reasoned, then he could confirm, as Franklin had postulated from his kite experiment, that lightning was indeed electricity. Galvani hung frog legs from brass hooks against an iron latticework to test his premise. When thunderstorms came by, the legs twitched. But something else happened: They also twitched when there were no thunderstorms. The twitching occurred, Galvani found, whenever the muscles came in contact with two different metals at the same time.

Galvani wasn't sure what the cause of this phenomenon was. Did the metals cause the twitching? Or did the muscles, even though dead, retain a sort of innate "animal electricity"? Maybe Galvani's interests in biology led him to lean toward the idea that the animal tissue of the frog legs possessed the electrical force he saw. But he published his results in 1791 and sparked a kind of revolution when another Italian, Alessandro Volta, saw his publication and began working on the problem.

Volta (1745–1827) read Galvani's *Commentaries*, repeated his experiments and tried another one on himself. He tried putting a piece of tinfoil and a silver coin in his mouth—one on top of his tongue, the other touching his tongue's lower surface, connecting them with a copper wire. He found that this rigging produced a distinctive sour taste in his mouth. He correctly surmised that this sour taste indicated the presence of an electrical charge.

"It is also worthy of note," he wrote, "that this taste lasts as long as the tin and silver are in contact with each other. . . . this shows that the flow of electricity from one place to another, is continuing without interruption."

The metals, he realized, were not just conductors—they were actually producing the electricity themselves! Galvani was wrong; the frog legs had exhibited not animal electricity, but metallic electricity. (But Galvani had played an important role, drawing attention by his experiment to a fact that would dramatically open up the door to the study of electricity, its use as a valuable tool in science, and the enormous industrial and commercial uses found for it in the 150 years since. His name has become a household word in expressions such as *galvanized by fear* and terms such as *galvanized iron* and *galvanometer*, an instrument designed to detect electric current.)

By 1797, Volta had succeeded in producing current electricity—not the static form of electricity of the Leyden jar, which had been the best available up to that time. And in 1800 he wrote to the Royal Society in London, describing the first battery, a continuous source of electric current.

DAVY'S ELECTROCHEMISTRY

Electricity held great fascination for everyone—in both scientific and social circles—at the end of the 18th century. Everyone was talking about Benjamin Franklin's experiments with a kite string and lightning, and socialites delighted in playing with static electricity at picnics and parties. But no one had succeeded in finding out much about what it was or how it worked, partly because no continuous source of it existed.

Not, that is, until Alessandro Volta invented what came to be known as the voltaic cell (see box). Volta's work not only opened up avenues for exploring the nature of electricity (producing spectacular results, both in theoretical physics and industry), but also provided chemistry with a breakthrough tool for discovering new elements and exploring the nature of chemical bonding.

And here our story doubles back to chemistry. No sooner had Volta communicated his findings to the Royal Society in London than another young scientist, Humphry Davy, began thinking about a way the voltaic cell could be used to solve some problems in chemistry. Davy, who is probably best known for his discovery of two elements, sodium and potassium, and for his invention of a safety lamp for miners, was hired in 1800 (along with Thomas Young, whose work we'll discuss in Chapter Four) to join the staff of the Royal Institution, a newly founded research laboratory and educational institution.

Davy was the oldest of five children, born in 1778, the son of a woodcarver in the town of Penzance on the western coast of Cornwall, England. In 1794, when young Davy was only 16, his father died, leaving the boy to support his family. So Humphry shortly began an apprenticeship with a local surgeon, but, by the time he was 19, he developed a keen interest in experimental chemistry and the boundary between chemistry and physics. He began to test the ideas in Antoine Lavoisier's *Traité élémentaire de chimie* (1789) and came to some revolutionary conclusions for the time. Based on observations he made while rubbing blocks of ice together, he asserted that heat was not an "imponderable fluid," as most chemists of the time thought, but a form of motion. Unfortunately, Davy was young and a little reckless, and he spoke with more confidence than his experiments warranted. As a result, the scientific community greeted his announcement rather coolly, with a large dose of skepticism, and Davy was deeply disappointed.

But in 1798 Davy became an assistant to Thomas Beddoes, a versatile chemist and physician specializing in the therapeutic uses of gases. At Beddoes's Pneumatic Institution in Bristol, Davy experimented, using himself as a guinea pig. He discovered how to prepare nitrous oxide (sometimes called laughing gas, still used widely by dentists), of which he breathed some 16 quarts all in one day, an experience he later said "completely intoxicated" him. He investigated the physiological effects of the gas and wrote a well-reasoned paper on it in 1800, which succeeded in establishing his reputation as a chemist (and attained notoriety among several figures in society, including the poets Coleridge and Wordsworth, who enjoyed visiting his lab to experiment with the intoxicating effects of his discovery).

Davy's scientific paper on nitrous oxide caught the eye of Count Rumford, a colorful American-born figure, whose work on heat as motion also had stirred up considerable controversy in the preceding decade, at a time when most chemists and physicists thought heat was caused by an imponderable fluid (that is, a fluid having no weight) called "caloric." Though working at the time for the government of Bavaria, Rumford, originally known as Benjamin Thompson, had come up with the idea of founding a Royal Institution in Britain to popularize science and apply its discoveries to everyday life, the arts and manufacturing. Rumford hired Davy as its first director of the laboratory, a great break for the aspiring young chemist.

The year was 1800. Before he left Bristol for London Davy had established to his own satisfaction that Volta's cell was producing electricity through chemical reaction, and he was quick to surmise that the reverse might also be true: that the use of electricity on compounds and mixtures might in turn produce chemical reactions.

However, over the next few years, his duties in London at the Royal Institution took him away from the subject. To raise money, the institution developed a highly popular lecture series, and Davy's charisma and enthusiasm took him far as one of the best lecturers of his day. (His spectacular demonstrations with electricity and amusing exhibitions of the "highs" produced by nitrous oxide didn't hurt.) Probably as much to keep a firm financial footing as to pursue the ideals of popularization, the institution focused on agricultural science, tanning and mineralogy, and Davy's several excellent papers on these subjects not only added to the stature of the institution but also enhanced his own.

But in 1806 he saw his opportunity. In the space of five weeks he performed 108 experiments in electrolysis, the use of electricity to produce chemical changes. In one brilliant coup that year, in a lecture "On Some Chemical Agencies of Electricity" to the prestigious Royal Society, Davy established theoretical links between electrolysis and voltaic action, and he gave one of the first explanations—certainly the first important one—of the electrical nature of chemical reaction. Substances combine chemically, he said, because of a mutual electrical attraction between atoms.

Davy also thought that electricity might be used to break the bonds between parts of compounds to isolate elements not yet discovered. Scientists had been working on several substances for years—lime, magnesia, potash and others—that seemed to be oxides of metals. But no amount of heat, or any other method anyone could think of, had ever succeeded in

Humphry Davy, founder of electrochemistry   (AIP Emilio Segrè Visual Archives)

breaking the tightly held oxygen away. At the end of his presentation in 1806 he prophetically mentioned his hope "that the new mode of analysis may lead us to the discovery of the *true* elements of bodies."

To try this trick, Davy built a huge battery, composed of more than 250 metal plates and much more powerful than Volta's little piles of metal disks and cardboard. The following year, working with a lump of very slightly dampened potash (a substance formed by soaking the ashes of burnt plants in pots of water), he attached an insulated metal electrode from the negative side of a battery to one surface of the lump. To another surface of the potash, he attached a metal wire running to the positive side of the battery, which, he noted, "was in an intense state of activity." At both points of contact, the potash began to fuse, giving off a gas on the surface that was attached to the positive pole. At the other contact point, no gas was given off, but "small globules having a high metallic lustre" began to form. They looked a lot like droplets of mercury and some of them burned brightly and exploded. Davy knew immediately that he had discovered a new element, which he called potassium (after "potash"). As his brother, John, wrote describing the experiment, when Humphry Davy "saw the minute globules of potassium burst through the crust of potash, and take fire as they entered the atmosphere, he could not contain his joy—he actually bounded about the room in ecstatic delight; and some little time was required for him to compose himself sufficiently to continue the experiment."

A few days later Davy used the same process on soda (now known as sodium hydroxide) and discovered sodium. The trick was working. Meanwhile, in Stockholm, Jöns Jacob Berzelius and his colleagues were pursuing similar experiments, and communications flew back and forth. Berzelius had found that he got an "amalgam," or alloy of some other metal with mercury, when he ran a current through a mercury compound added to lime or baryta. That gave Davy another key, and within a few months, by applying a strong heat to the amalgams Berzelius described (and to others), he also isolated magnesium (from magnesia), calcium (named after "calx"), strontium (from a mineral named for a Scottish town called Strontian), and barium (from baryta). Davy was racking up an impressive list of discoveries.

To Davy also goes the credit for testing a green gas called oxymuriatic acid and recognizing it in 1810 as an element that he called chlorine (for its greenish color).

The year 1812 was a watershed for Davy, the year in which he published his *Elements of Chemical Philosophy*. He followed up quickly on that with the more applied *Elements of Agricultural Chemistry*. He was knighted, in recognition of his accomplishments, in April 1812 and shortly thereafter married a wealthy Scottish widow, Jane Apreece. In 1813 he resigned his position as professor at the Royal Institution and set off for Europe with his new wife and a young assistant he had recently taken on, Michael Faraday, whose story figures large later in the century. Despite the fact that England was at war with France at the time, as Davy remarked, "There is never a war among men of science," and Napoleon welcomed Davy's visit, during which Davy and Faraday called on many of the prominent scientists on the Continent. For Faraday the trip was a stunning introduction to the cutting edge of science.

Davy became president of the Royal Society in 1820 and began working on a means to prevent corrosion of the copper sheathings on ship bottoms, but he became ill and spent most of his time after 1823 in Switzerland, where he died at the age of 51. There his stature was so great that he was given a state funeral.

The year was 1829, and for chemistry, the century had only just begun, bolstered by the atomic theory of John Dalton, the new tool invented by Volta and the extraordinary discoveries of new elements made by Davy and others. New challenges lay ahead: to find order in the chaos of new elements, to continue to search out more new elements and to make sense of the vast jungle of molecules that form with the element carbon. Progress in all these fields was soon to come.

C H A P T E R   2

CHEMISTRY'S ROLLERCOASTER OF COMPLEXITY AND ORDER

*B*y 1830, the number of known elements had swelled to 54. In addition to the six discovered by Davy between 1807 and 1808, 10 more had turned up, including boron, iodine, lithium, silicon, bromine and aluminum. Obviously more than "a few simple elements" made up the stuff of the universe, and instead of the convergency the century had seemed to promise, confusion now seemed to reign in chemistry.

For one thing, no one used the same symbols to mean the same thing. Many strange, mysterious signs still persisted, borrowed long ago by the alchemists from astrology. For gold, the symbol was a circle with a dot in the center, for silver a crescent. The symbol for sulfur was a triangle pointing upward, and for antimony a little crown. Most of these made no real sense to anyone. Dalton had offered a system that used a circle differentiated in some way for every element. But even this was difficult to remember. Finally, in 1826, Berzelius came up with the simple idea of using the first letter of each element's name as its symbol. O was oxygen, N was nitrogen, S was sulfur and so on. When an initial letter was already taken, the next distinguishing letter was added. So calcium was Ca and chlorine was Cl. This system is still in use today. Some confusion still existed between languages: German chemists called one element Stickstoff, while the French called it azote and the English called it nitrogen. So Berzelius settled on the latinized names as his sources, and the symbols were adopted internationally. Luckily for those whose language is English, most elements were already known by their latinized names, with a few exceptions such as gold (Au from aurum), silver (Ag from argentum) and sodium (Na from natrium).

Friedrich Kekulé von Stradonitz [KAY koo lay] also came up with an idea for arranging atomic symbols in structured diagrams, representing the arrangement of atoms in molecules. In Kekulé's system, water (H_2O), for example, became H—O—H. Likewise, the three hydrogen atoms of ammonia (NH_3) clustered in a triangle around the lone nitrogen atom.

$$H$$
$$|$$
$$H - N - H$$

And soon Kekulé structures caught on.

But controversy reigned about the formulas for even the most common compounds. No one could agree about the atomic weights of the various elements and many confused atoms with molecules in writing formulas. For even such a common compound as acetic acid (vinegar), various factions of chemists touted as many as 19 different formulas.

THE KARLSRUHE CONFERENCE

Something had to be done. At the heart of the movement for greater clarity stood Kekulé, who called the first international scientific conference to try to clean up the mess in chemistry. The First International Chemical Conference, as it was called, took place in 1860 in the small town of Karlsruhe, Germany, on the banks of the Rhine River, just across the border from France. A total of 140 delegates attended, including most of the prominent chemists of the day.

But this was a group of adamant and opinionated scientists, and the conference got off to a stormy start, with little agreement on anything, certainly not on atomic weights. Then Stanislao Cannizzaro took the floor.

Cannizzaro was a fiery, combative individual who was accustomed to conflict. In fact, in 1848 he had fled to France from his native Sicily to avoid punishment for fighting on the losing side of a Sicilian insurrection against the then-ruling government of Naples. In France, he'd been giving considerable thought to the mess chemistry was in, and in 1858 he had published a paper resurrecting Avogadro's hypothesis, which no one had thought about for almost 50 years—the idea that equal volumes of different gases (at the same temperature) must contain the same number of particles. He arrived in Karlsruhe prepared with a fiery defense of atomic weight, Avogadro's hypothesis and a clear distinction between atoms and molecules. Use Avogadro's hypothesis to determine the molecular weight of gases, he said; use Gay-Lussac's law of combining volumes; and use Berzelius's atomic weights. With this combination, Cannizzaro maintained, many of the problems would be resolved. He backed up his speech with pamphlets, convincing some at the conference and many more shortly thereafter. One, in particular, would go back to Russia and think about it a great deal.

MENDELEYEV'S SOLITAIRE

With his long hair flying in the wind, graying beard, and commanding posture, Dmitry Mendeleyev [men deh LAY ef] looked more like a biblical prophet than a patriarch of the sciences the day he singlehandedly piloted his tiny basket high into the skies beneath a giant balloon. The year was 1887, and he wanted to photograph a solar eclipse from the closest, best vantage point he could get. That meant a balloon, a one-person craft. So, not to be stopped at the brink of such an opportunity, Mendeleyev took off, took his photographs and landed, even though he didn't know the first thing about flying such a contraption. He was a flamboyant individual of principle and courage, unafraid of skeptics, naysayers, political pressures or airborne craft. A native of Siberia, he was the magician who, some 18 years earlier, had brought order to the chaotic mess of elements chemists had by that time discovered. He was also the first scientist from the Russian empire whose work made a timely impact in Europe, and in 1955, nearly 50 years after his

Dmitry Mendeleyev, whose periodic table of the elements helped chemists see inherent order where there seemed to be chaos
(AIP Emilio Segrè Visual Archives, W. F. Meggers Collection)

23

death, his extraordinary contributions to chemistry and physics received the perfect tribute: a newly discovered element, mendelevium, named in his honor.

Possibly of Mongolian descent on his mother's side, Mendeleyev (1834–1907) was the youngest of a very large family of some 16 or 17 children, grandson of the first newspaper publisher in Siberia and son of the high school principal and a resourceful mother who ran a glass factory. He learned science as a child from a political prisoner who had been banished to Siberia as punishment. But Mendeleyev's father died while he was in his early teens and his mother's glass factory burned down soon after her husband's death. So, with most of her children grown, his mother set out in 1849 for the big cities of Russia to gain admittance to college for her youngest. She succeeded in St. Petersburg, where Mendeleyev was admitted to the university with the help of a friend of his father.

After graduating in 1855, Mendeleyev set out for France and Germany in 1859 for graduate studies in chemistry. While there he worked with Robert Wilhelm Bunsen (who invented the Bunsen burner) and attended the First International Chemical Conference at Karlsruhe. The strong arguments Cannizzaro had made there concerning atomic weights continued to press on Mendeleyev's mind when he returned to St. Petersburg, where he began teaching at the university in 1861 and received an appointment as professor of technical chemistry in 1866.

Some scientists had speculated that closeness in atomic weight might be related to similarities in characteristics among the elements. Cobalt and nickle, for example, had atomic weights so close that most chemists at the time couldn't differentiate them, and their characteristics were very similar. But the hypothesis didn't hold up. Take chlorine and sulfur, with atomic weights of approximately 35.5 and 32 respectively. One is a yellow solid, the other a green gas—strikingly different. So they began to look for other relationships. By 1861, a number of chemists had been playing around for several years with the idea of triads or small groups of elements that seemed to be "families," based on similarities in their properties. As early as 1817 Johann Wolfgang Döbereiner had begun to notice some consistent relationships in the atomic weights of certain groups of similar elements—with the middle member's atomic weight equal to the mean of the other two atomic weights. For example, in the triad of calcium, strontium and barium, the atomic weight of strontium (figured at the time as 88) was roughly the mean of calcium's (40) and barium's (137). Likewise, the melting point of strontium (800° C) lay between that of calcium (851° C) and barium (710° C). And calcium is considered active in chemical reactions, with barium more so and strontium—in between! The list of ways that strontium was "in between" calcium and barium went on and on. Such triad relationships were intriguing, and other scientists added to the idea.

In 1864, the London industrial chemist John Alexander Reina Newlands (1837–98) was the first to notice that a table of elements arranged by order of atomic weights showed a pattern in which "the eighth element, starting from a given one, is a kind of repetition of the first, like the eighth note in an octave of music." He called this discovery the "law of octaves," but he was virtually laughed out of the meeting of chemists at which he announced his ideas. Why not just alphabetize the elements to see what patterns you get? jibed George Carey Foster, a professor of physics. Newlands's table of elements did have some flaws, but he had in fact recognized a useful pattern. George Carey Foster, though a competent physicist, is known primarily for his taunting remark—illustrating that a scientific idea that may seem to go nowhere today can lead to new insights tomorrow, and ill-considered scoffing can come back to haunt the scoffer. More than 20 years later, the Royal Society awarded Newlands the Davy medal for his work, which also included organizing the elements in order of ascending atomic weight and assigning a number to each according to its position in his table.

But Dmitry Mendeleyev was the one who played with the idea of order among the elements the most creatively and pushed it to its most logical conclusions. Mendeleyev was fond of a type of solitaire game called patience. So he played patience with all the known elements, their symbols, atomic weights and unique properties marked on cards. Then he started arranging them in groups. And he found that, if you lined the elements up in the order of increasing atomic weight, similar characteristics would occur every so often—periodically spaced. For example, he found that hydrogen (with an atomic weight of 1 and number one on his list), fluorine (9th on his list) and chlorine (in the 17th position) were spaced 8 spaces apart, like Newlands's "octaves" and shared similar properties. He tried listing all the groups with similar properties in the same vertical column. And he began to work out a table, with atomic weight increasing from upper left to lower right, and families of elements stacked in columns.

But Mendeleyev's great daring was that, where elements wouldn't fit into his scheme for the table, he played the game as he would a game of solitaire, recognizing that he might not have all the cards in his hand—some cards might still be in the deck. So, if a slot called for an element with certain properties, and there was no such element as far as anyone knew, then he left gaps in his table for the elements still in the deck—those not yet discovered. What's more, he even named some of them: eka-boron, eka-aluminum and eka-silicon. Eka-aluminum was the gap in the column below aluminum, eka-silicon was the one below silicon. And he predicted their properties. His work, published in 1869, was translated immediately from Russian into German (in this he was far luckier than any Russian scientist before him, since Russian work was usually lost to the rest of the world for years because it went untranslated). But in Europe, everyone thought he was crazy, possibly even wrote him off as a Russian mystic.

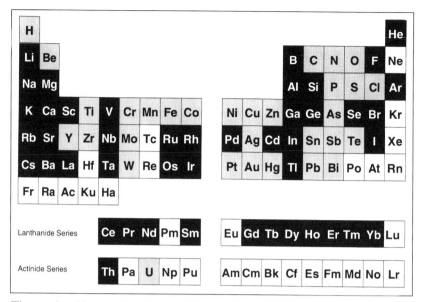

The periodic table as it looks today. The elements that are shaded gray were familiar to scientists at the beginning of the 19th century; those shaded black were discovered between 1800 and 1895. The most recent additions shown here, from Americium (Am) to Lawrencium (Lr) in the Actinide Series and the unofficially named elements Kurchatovium (Ku) and Hahnium (Ha), are part of a series of synthetic elements added since 1944 that extend the known list of elements as high as 112. (This chart, like most, shows only the 105 most accepted.) See p. 111 for a key to the symbols.

THUMBPRINTING THE ELEMENTS

Meanwhile, as Mendeleyev was working on his periodic table, a wonderful new tool came on the scene: the spectroscope. It would prove useful not only for chemists but also for astronomers and physicists.

The idea began early in the century with the work of a young optician named Joseph von Fraunhofer. The son of a glazier, he was orphaned at 11 and became the apprentice of an optician. One baleful day the entire building he lived in collapsed around him, and he was the sole occupant to survive. But he was in luck. The elector of Bavaria, Maximilian I, heard the tragic story and gave him enough money to start his own career as an optician.

Fraunhofer developed an international reputation for high quality and precision in his work, and his prisms and optical instruments were used by several prominent astronomers. In 1814, as he tested some lenses he was making, he made use of a prism, which, as Newton had shown a century earlier, could break the Sun's white light into the colors of the spectrum. As he did so he noticed some strange black lines that seemed to punctuate the solar spectrum—in fact, he saw at least 600, some wider, some narrower, dividing portions of the spectrum. (Newton, using considerably less accurately ground prisms, probably could not see them because the imperfections of the glass would have made them fuzzy.)

Each color of the spectrum, Fraunhofer knew, correlates with a unique wavelength of light. (Light waves have crests and troughs, much like the waves of the ocean, and the measurement from one crest to the next is called a wavelength.) Shorter wavelengths fell at the red end of the spectrum, while longer ones fell at the violet end. Fraunhofer noticed the position of the prominent lines in the spectrum, labeled them A through K (which they are still labeled today), figured out their wavelength, and observed that their positions in the spectrum always remained the same. The strange black lines seemed to be a kind of code. Definitely they had some significance. He tried using different light sources—the direct light of the Sun and the reflected light of the Moon and planets. Then, even starlight. Each different star, he found, seemed to leave a different code, a different thumbprint. But no one could crack the code, and Fraunhofer died in 1826 of tuberculosis at the age of 39 without ever finding out the meaning of the "Fraunhofer lines" named after him.

Half a century later a team of physicists at the University of Heidelberg— Gustav Kirchhoff and Robert Wilhelm Bunsen—developed an instrument they called a spectroscope, which passed light through a narrow slit before passing it through a prism. The slit controlled the source of the light, and, as a result, different wavelengths were displayed differently and, when viewed against a scale, became much easier to differentiate and interpret.

Using the special burner devised by Bunsen, which gave off very little light itself, Kirchhoff and Bunsen heated various chemicals to incandescence (the heat at which they gave off light) and noticed that each chemical gave off its own distinctive pattern of colored lines. Sodium vapor, for example, when brought to a glowing heat, produced a double yellow line: its thumbprint. Once the thumbprints of all the elements were known, any ore or compound—any substance, in fact—could be heated and its components could be analyzed in this way. What's more, the spectroscope could fingerprint extraordinarily tiny amounts of an element.

Kirchhoff and Bunsen first announced their invention publicly on October 27, 1859, and the spectroscope, inevitably, began to uncover new elements, one after another. Cesium, named after its distinctive blue spectral line, was discovered on May 10, 1860. Rubidium, named after the red line that tipped off its existence, was discovered the following year. A new run of elements had begun to flow.

Then in 1875, a French chemist named Paul Emile Lecoq de Boisbaudran [luh KOHK duh bwah boh DRAHN] found a spectral line he had never seen before in a hunk of zinc ore from the Pyrenees mountains. One of the first to enter the exciting new field in 1859, after searching with his spectroscope for 16 years he at last had found a new element. He called it gallium, after the Latin name for France (or, maybe, after himself, since *Lecoq* means "rooster" in French, which is *gallus* in Latin). When Mendeleyev read the description of the new element, he was elated. Gallium had almost exactly the same properties he had predicted for eka-aluminum! The new element slipped easily into its place on the periodic table, and suddenly everyone began to take Mendeleyev more seriously. The powerful tool of spectroscopy had won the day.

Another element, scandium (named for Scandinavia), discovered in 1879, fit almost perfectly in the place Mendeleyev had left for eka-boron. And in 1886, when the element germanium (named for Germany) was discovered, it filled the eka-silicon spot. For most people Mendeleyev's periodic law had finally gained acceptance. He had recognized a natural order, in the manner of every good scientist, where chaos had seemed to reign.

But no one knew why this order, this periodicity, existed. That required knowing about the nucleus of the atom and its structure, and in the 19th century scientists were not yet ready to part with the idea that the atoms with which they were working were unsplittable. Yet, the way the number of elements kept steadily increasing, chemists seemed to be getting farther and farther away from the few simple building blocks of nature they had set out to find. The number would soon climb to over 90. (And the 20th century would see that number increase, with the many new elements created by nuclear chemists, to 112, and still counting.)

In the last five years of the century Robert John Strutt, better known as the famous English physicist Baron Rayleigh, and his assistant, a Scottish chemist named William (later Sir William) Ramsay, repeated an experiment performed a hundred years earlier by Henry Cavendish, this time using the spectroscope. As a result, they discovered argon, and Ramsay went on in the following years to discover helium and, with Morris Travers, the inert (completely unreactive) gases neon, krypton and xenon. For these, no slot remained in Mendeleyev's table. Could this be its downfall? But no, the answer was simple: The great Siberian card player was not clairvoyant; he had left out one entire column on the right-hand side of his table, which is where those elements reside today.

BIRTH OF ORGANIC CHEMISTRY

Meanwhile, as Dalton, Davy and Mendeleyev succeeded in revolutionizing inorganic chemistry, another, even more confusing, field was also undergoing a major transformation. In 1807 Berzelius named the class of chemicals that originated in living things organic, and gave the name inorganic to those that did not. Organic substances, he maintained, functioned by completely different laws than their inorganic cousins and appeared in many ways to be vastly different. Most scientists, including Berzelius, assumed this difference came from the presence of some "vital" force that linked organic chemicals to the living or once-living matter in which they were found, or by which they were produced. And no one had ever created an organic compound from inorganic substances. And, according to Berzelius, no one ever would.

Then one day in 1828 Friedrich Wöhler [VOH ler], a student of Berzelius, was working in his laboratory on some problem having to do with cyanides, when he heated a quantity of ammonium cyanate. He was stunned to see the results: He had produced a compound that looked exactly like urea. This seemed highly unlikely, since urea, a component of urine, is the primary nitrogenous waste of mammals, unquestionably organic. Skeptical, Wöhler tested the substance he had produced. It was definitely urea. And on February 22, 1828, he announced to Berzelius that he had produced an organic compound out of an inorganic compound.

Berzelius, never easy to convince, made a case for the idea that ammonium cyanate can be seen as organic, not inorganic. So Wöhler's discovery may have been a bit unsure—but other chemists were challenged by his achievement to try other inorganic compounds and found that, in fact, organic compounds can be synthesized from inorganic materials. Then in 1845 Adolph Wilhelm Hermann Kolbe succeeded for the first time in synthesizing an organic compound (acetic acid) directly from chemical elements. Perhaps no such thing as a "vital" force existed after all.

But, if not, then why did Jean-Baptiste Biot discover in 1815 that when he produced tartaric acid in the laboratory, it failed to polarize light (where the transverse vibrations of the light waves are confined to just one direction), when tartaric acid produced by grapes did polarize light? The two batches of acid had the same components in the same ratio, the same chemical formula. More such pairs were found by Justus von Liebig and Wöhler in the 1820s. And in 1830, Berzelius, the great namer, gave the name *isomer* to pairs of compounds that had the same chemical formulas but behaved differently. "Organic chemistry," wrote Wöhler in perplexity to Berzelius in 1835, "appears to me like a primeval forest of the tropics, full of the most remarkable things."

Louis Pasteur [pas TUHR], whom we'll meet again later in this book, did his first serious work in chemistry on the strange problem that Biot had found with the tartaric acid isomers. Pasteur tried isolating single crystals of the laboratory-produced isomer and found that they in fact did polarize light. Some polarized in one direction, and others in the opposite direction. By 1844 he had an answer. The two types of crystals canceled each other out in the substance made in the laboratory, which is why it appeared that the whole substance did not polarize light.

EXPLOSIVES, DYES, PERFUMES AND PLASTICS: ORGANIC GIFTS TO INDUSTRY

From the unlikely raw materials of coal, water and air, several lucrative chemical industries emerged in the 19th century: explosives, dyes, perfumes and plastics.

The first synthetic explosive, nitrocellulose, was discovered completely by accident by Christian Schönbein in 1846. He'd been working one day in his lab and used his wife's apron to wipe up some spilled chemicals—probably sulfuric and nitric acids. The apron—whose cellulose fibers combined dramatically with the acids—suddenly exploded and disappeared. Also known as guncotton, nitrocellulose caused many deaths by premature explosions in the early years of its use.

A cousin of nitrocellulose known as nitroglycerine was also discovered in 1846. Highly volatile and unstable, it was used in tunneling and blasting, also sometimes with disastrous results. Ways were found to tame both substances so they could be used more safely, and the results are cordite and dynamite, respectively. The use of such modern explosives has transformed the construction of big engineering projects such as highways, bridges, tunnels and dams, as well as mining.

Another chemical industry was accidentally begun by an English chemist named William Perkin a decade later, in 1856, with the discovery of a dye made from aniline that produced the color mauve. Actually, he was trying to make synthetic quinine (used to treat malaria), but mauve dye soon made him wealthy. Aniline, Perkin found, was not available on the open market; so he manufactured it from benzene, whose structure Kekulé

Kekulé's structural formulas, meanwhile, helped explain what was going on in some of these complex organic compounds, some of which bonded with double and triple bonds, which Kekulé's system could show with double and triple dashes. Isomers could have the same atoms in the same ratio, but arranged differently. For example, ethyl alcohol can be represented like this:

$$
\begin{array}{ccc}
\text{H} & \text{H} & \\
| & | & \\
\text{H}-\text{C}-\text{C}-\text{O}-\text{H} \\
| & | & \\
\text{H} & \text{H} &
\end{array}
$$

And dimethyl ether, which has the same number of hydrogen, carbon and oxygen atoms, can be shown like this:

would soon unlock. His colleague and teacher, the German August Wilhelm von Hofmann, discovered how to make a magenta dye the following year, and Germany soon became the seat of a very profitable chemical dye industry. An orange-red crystalline named alizarin was synthesized in Germany by Karl Graebe in 1868, followed by indigo, synthesized by Adolf von Baeyer in 1880. (In a sort of side benefit to science, biologists soon found that certain plant and, especially, animal cells were easier to see under a microscope when stained with some of these dyes.) Graebe contributed to structural understanding of organic molecules by extending Kekulé's benzene ring to the structure of naphthalene, and Baeyer formulated the structure of indigo in 1883.

In 1875, Perkin scored again by producing the first synthesized perfume ingredient, coumarin, and from this discovery, another great industry grew.

Meanwhile, plastics got their start in the 19th century as well, with the synthesis of celluloid. The English chemist Alexander Parkes first converted the explosive nitrocellulose to a useful nonexplosive (though flammable) substance in 1865. Soon afterward, the American inventor John Wesley Hyatt was looking for a better billiard ball in an age when they were made of ivory, and he improved on Parkes's celluloid for the purpose. English and American chemists dominate the plastics field to this day, and a wide array of types have been invented in the 20th century. The varieties seem endless, ranging from textiles such as rayon, nylon and polyester to moldable, solid plastics—sometimes pliable, sometimes rigid—used for everything from plumbing pipes to toothbrushes, and from drinking straws to shower curtains.

$$H - \underset{\underset{H}{|}}{\overset{\overset{H}{|}}{C}} - O - \underset{\underset{H}{|}}{\overset{\overset{H}{|}}{C}} - H$$

Carbon atoms, Kekulé pointed out in 1858, could combine directly with each other (unlike most other atoms), forming lengthy and complex chains. Also, he said, carbon atoms have a valency of four—that is, they always combine with exactly four other atoms. And he made clear that, by studying the products of a reaction, one could draw conclusions about the molecular structure of an organic molecule.

In 1861, Kekulé published the first volume of his textbook on organic chemistry. In it, he cut through a long-term, entangled controversy with a single, clean stroke. He defined organic molecules simply as those that contained carbon and inorganic molecules as those that did not. He made no reference to the issue of living or once-living substances. It was a blow against the idea that organic molecules somehow contained an indefinable "vital force," and it offered a new and useful way of looking at the field of organic chemistry.

GRABBING THE RING

For organic chemistry another problem still remained. No one had yet been able to explain the structure of benzene (C_6H_6), a coal tar product discovered by Michael Faraday in 1825. Of course, even without knowing the structure of benzene, William Perkin (see box pp. 30–31) and others working in the synthesis of dyes had been making progress. But no one could show how these atoms could fit together with each other in a way that explained how this molecule typically combined.

Then, one day in 1865, as Kekulé later wrote:

> I was sitting, writing at my text-book; but the work did not progress; my thoughts were elsewhere. I turned my chair to the fire and dozed. Again the atoms were gambolling before my eyes. This time the smaller groups kept modestly in the background. My mental eye, rendered more acute by repeated visions of this kind, could now distinguish larger structures, of manifold conformation: long rows, sometimes more closely fitting together; all twining and twisting in snake-like motion. But look! What was that? One of the snakes had seized hold of its own tail, and the form whirled mockingly before my eyes. As if by a flash of lightning I awoke; and this time also I spent the rest of the night working out the consequences of the hypothesis.

What Kekulé came up with was what we now call the benzene ring, a molecule composed of benzene's carbon and hydrogen molecules arranged not in an open chain, but in a closed hexagon, with alternate single and double bonds in rapid oscillation.

A Dutch chemist, Jacobus Van't Hoff, translated many of Kekulé's structural ideas into three-dimensional models that served to clarify much of organic chemistry, including the isomer puzzle that Biot and Pasteur had worked on. Kekulé's structural insights brought order to organic chemistry where at the beginning of the century incredible confusion had existed, and though many theoretical refinements have been made since, his ideas still serve to guide chemists through the thicket of synthesis and provide a model that visualizes the organic molecule and predicts its reactions.

For chemistry, the 19th century was a time of extraordinary productivity. Two important new tools, electricity and spectroscopy, gave chemists new

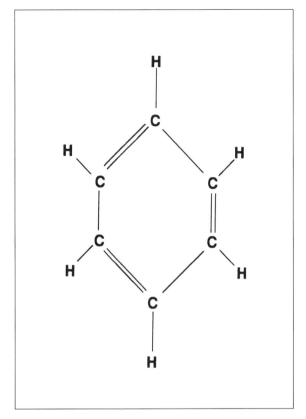

Kekulé's benzene ring

ways to manipulate and observe materials and transformed their science in the same way that the telescope had done for astronomy and the microscope for biology. The number of known elements nearly doubled. Mendeleyev's periodic table began to make sense of them and provided a working matrix for the great breakthroughs yet to come in both chemistry and physics at the turn of the century and in the early 1900s. The birth of organic chemistry opened up enormous industrial potential for applied chemistry, including the invention of new dyes and materials.

Most important, at the outset, the birth (or rather, rebirth) of atomic theory enabled Dalton, Avogadro and those who followed to begin to make sense out of the rules of chemistry—how substances reacted and bonded with each other—as well as the properties of gases.

Of course, not everyone was happy with atomism, either at the time Dalton first set it forth, or even later, at the end of the century. The highly influential physicist Ernst Mach (1838–1916) opposed atomism right up to the time of his death. It was one thing, he said, to observe that two volumes of hydrogen gas combined with one volume of oxygen gas to form water vapor; it was quite another to postulate that two atoms of hydrogen, which could not be seen, combined with an invisible atom of oxygen to form a molecule of water, which also could not be seen. But atomism at the very least, most scientists conceded, provided an excellent model, which, with the use of notations symbolizing atoms and their inter-reactions, made discussions of chemistry much clearer.

Atomism also opened the way to one of the great key discoveries of the century: an understanding of the nature of heat and thermodynamics, an area that had been clouded by mystery for centuries.

INDESTRUCTIBLE ENERGY

*T*wo great forces turned wheels and intrigued minds of the 19th century: steam and electricity. At the start of the century all industry had been transformed by James Watt's steam engine, which also provided inspiration for the theoretical study of energy. By mid-century, transportation was also transformed, with all the major ports of England linked by steam railroads and some 30,000 miles of track crisscrossing the North American continent. And by the end of the century, electricity had begun to light up the world and provide the productive power for industry.

Scientists who delved into the heart of these two great sources of power turned up a treasure trove of insights into nature that fed back into the technological development of western Europe, the British Isles, North America and the entire world. The key, as Joseph Black and James Watt discovered in the previous century, was understanding heat, its nature, its behavior, and, most of all, thermodynamics—the study of how thermal energy converts to other forms of energy and vice versa.

EARLY WORK

For most 18th-century chemists and physicists, heat was an invisible, "imponderable" (that is, weightless) fluid called "caloric." When ice melted, it lost caloric. When water froze, it gained caloric in a sort of chemical reaction between water and heat. The theory, sometimes known as the material theory of heat, seemed to work well as an explanation: If you place a hot object next to a cold one, the heat does seem to flow from one to the other, as if it were a fluid. And, matter expands when heated, as if some fluid were being added to it. Caloric seemed to make sense, so very few scientists saw any reason to challenge it.

Few, that is, except the American-born Bavarian elector Count Rumford (who left America to avoid prosecution for collaborating with the British during the Revolutionary War). He also picked up recruits among the younger generation of English scientists around 1800—including Humphry Davy and Thomas Young. Rumford had realized that the act of boring cannons with dull instruments should have produced less heat (releasing less caloric) than boring with sharp instruments; use of the sharp instruments should have released more caloric, since they abraded the material more effectively. In fact, exactly the opposite was true. To explain this, Rumford suggested that heat must be a kind of motion. It was not an idea that caught on quickly.

But as the 19th century dawned, John Dalton's atomic theory began to add credence to the idea that there might be tiny, invisible particles capable of agitating in a balloon full of gas or a vat of water or a block of ice—moving faster if hot, slower if cold.

An idea along these lines, known as kinetic theory (*kinetic* means produced by motion), had been introduced by Daniel Bernoulli in 1738, but atoms and molecules weren't really taken seriously at the time. A few other people also tried to propose it after Dalton, but they were unknowns and no one paid much attention.

French scientists, meanwhile, were working on the theoretical basis for Watt's steam engine. Watt, an engineer, and his Scottish and English scientist friends were practical, hands-on doers, many of them self-educated. The French, with their École Polytechnique in Paris, were stronger at theoretical science, with a preference for the material theory of heat (caloric). Jean-Baptiste Joseph Fourier (1768–1830), who had a strong influence on mathematical physics, published a paper, *Théorie analytique de la chaleur* ("Analytical Theory of Heat," 1822), in which he presented a new method for mathematical analysis and was the first to clearly state that a scientific equation must involve a consistent set of units, an idea known as "Fourier's theorem." He also examined the flow of heat through solids and a theory of dimensions that Descartes had suggested. But Fourier wasn't interested in the mechanical force associated with heat, and, in fact, he thought that "dynamical theories" and "natural philosophy" occupied two different and unrelated realms.

In Germany, meanwhile, the kinetic theory of heat was gaining ground. Friedrich Mohr (1806–79), a pupil of the chemist Justus von Liebig, wrote in 1837:

> *Besides the known fifty-four chemical elements there exists in nature only one agent more, and this is called force; it can under suitable conditions appear as motion, cohesion, electricity, light, heat, and magnets. . . . Heat is thus not a particular kind of matter, but an oscillatory motion of the smallest parts of bodies.*

All these ideas circled around one central thought but didn't quite land on it. It was up to an avid experimentalist named James Joule to give it quantitative value.

JOULE'S MEASUREMENT

James Prescott Joule (1818–89) was a fanatic about heat. He measured the heat of everything. Even on his honeymoon, he measured the water temperature at the top of a waterfall he and his wife were visiting and compared it to the temperature of the water at the bottom.

In a classic experiment that he performed in 1847, Joule inserted a paddle wheel into a tank of water of which he measured the temperature. Then he spun the paddle wheel for a long time, resulting in a very gradual rise in the temperature of the water. Joule measured the amount of work done by the paddle wheel and the rise in water temperature; that is, he figured out how

James Joule
(AIP Emilio Segrè
Visual Archives)

much mechanical energy produced how much heat, a value now known as the mechanical equivalent of heat. Joule spent 10 years or more of his life measuring the heat produced by every process he could think of—mechanical, electrical or magnetic—and in every medium he could think of.

Others had tried to come up with a figure for the mechanical equivalent of heat before him. Rumford had done it, but had come out way too high; Julius Robert Mayer 1814–78 also had figured it out, but not as accurately as Joule. Joule's was the best figure up to his time, and he backed it up with voluminous experimental data. In his honor a unit of work or energy, equal to 10 million ergs or about a quarter of a calorie, is named a joule.

His work led directly to the recognition of a fundamental principle known as the first law of thermodynamics, and for this he often shares the credit with the man who formulated it.

THE FIRST LAW

So to Lavoisier's principle of the indestructibility of matter, it was Hermann von Helmholtz (1821–94) who added a corollary law in 1847: "Nature as a whole possesses a store of energy which cannot in any wise be added to or subtracted from." The quantity of energy in the universe is as indestructible as the quantity of matter; matter cannot be created or destroyed, nor can energy. (Julius Robert Mayer had put forth a concept of the conservation of energy in 1842, before either Joule's work or Helmholtz's, but it was less completely supported by evidence than Helmholtz's.)

Known as the first law of thermodynamics, this idea is sometimes summed up as "You can't get something for nothing." Or, put still another way: You cannot get more energy out of a reaction than you put into it. Or:

thermal energy input = useful energy + waste energy

As Black and Watt saw, a heat engine (of which Watt's steam engine was the first successful example) could convert thermal energy stored in gases into the kinetic energy of turbines or pistons. That is, since gases expand when heated, thermal energy stored in steam could be converted to motion. The original source of energy in such a system is the chemical potential energy of the fuel—wood or coal—that is used to produce steam.

The first law of thermodynamics is one of the most revolutionary ideas in the history of physical science, and according to science historian A. C. Crombie, "Its implications and the problems it posed dominated physics in the period between the electromagnetic researches of Faraday and Maxwell and the introduction of the quantum theory by Planck in 1900." It would also prove to need extension to include both energy and matter together

Hermann von Helmholtz, one of the founders of the principle of conservation of energy, is also known for his contributions to ophthalmology, anatomy and physiology (Courtesy National Library of Medicine)

with the advent of Einsteinian physics in the 20th century, when it became evident that energy can sometimes be converted to matter and vice versa.

As James Clerk Maxwell wrote in a tribute to Helmholtz:

> *To appreciate the scientific value of Helmholtz's little essay on the Conservation of Force, we should have to ask those to whom we owe the greatest discoveries in thermodynamics and other branches of modern physics, how many times they have read it over, and how often during their researches they felt the weighty statements of Helmholtz acting on their minds like an irresistible driving power.*

In his later years, Helmholtz became a mentor of Max Planck (1858–1947), the founder of quantum theory, and through Planck, Helmholtz's influence would also reach far forward into the 20th century.

THE SECOND LAW

Nicolas Léonard Sadi Carnot (1796–1832), unlike Fourier, was a French engineer, and his approach was more practical. He likened the steam engine to a water wheel—a somewhat flawed analogy—and he initially put forth the idea that the boiler in a steam engine put out the same amount of heat as was received by the condenser at the lower temperature, that no heat was lost. While this is not so, Carnot made the important link between the heat of the fire, the pressure of the steam and the mechanical motion (or work) of the engine. He recognized that the energy output of an engine depended on the difference between the high temperature at the boiler and the lower temperature at the condenser and the amount of heat that passed between them. And he speculated that the total energy of the universe was constant, and only changed from one form to another. But Carnot died of cholera at the age of 36 before he had a chance to develop his ideas any further. His ideas were published in 1824 in his only work, *On the Motive Power of Fire*, but it had considerable influence on those who followed.

German physicist Rudolf Clausius (1822–88) was not an experimentalist; his great gift lay in the ability to interpret and perform mathematical analysis of other scientists' results. Clausius came to the conclusion in 1850 that heat could not by itself pass from one body to another at a higher temperature. Considered to be another of the 19th century's

Rudolf Clausius, who, with Lord Kelvin, cofounded the second law of thermo- dynamics (AIP Emilio Segrè Visual Archives, Physics Today Collection)

William Thomson, Lord Kelvin (The Smithsonian Institution, Courtesy AIP Emilio Segrè Visual Archives)

major discoveries in physics, his formulation has become known as the second law of thermodynamics.

Irish-born William Thomson (1824–1907) became known as Lord Kelvin of Largs in Scotland and is often referred to by either name. In addition to contributing to the dynamical theory of heat, he synthesized the ideas of Carnot and Joule in a paper published in 1851 on the convertibility of heat into mechanical energy, also a version of the second law of thermodynamics. For this contribution he is sometimes given some of the credit, alongside Clausius, for discovering the principle.

The second law of thermodynamics can be stated simply: You can't break even. Suppose a diver stands at the top of a cliff over a deep pool. At this moment the diver has gravitational potential energy. When he or she jumps, that energy is transformed to kinetic energy, which, in turn, is transferred to the water as thermal energy when the diver hits it. This process, by the way, won't reverse spontaneously (at least, not usually); energy transformations have a preferred direction. Although it would be possible to see the diver bounce back up to the top of the cliff, some kind of bungee cord or spring or a crane would be required to achieve the feat. Otherwise the diver will have to hike back up or hitch a ride in a dune buggy. Or, in another example, hot soup always becomes cold spontaneously, but cold soup doesn't become hot unless heat is applied from an outside source.

Another way of expressing the second law of thermodynamics is to say that in a closed system—one having no outside source of energy—entropy increases. Entropy is a measure of the disorder of a system: the greater the disorder, the higher the entropy. And, because entropy tends to increase, thermal energy doesn't flow from cooler to warmer (molecules and atoms are more ordered in cooler solids than in warmer liquids and gases), and, in general, natural processes move toward greater disorder.

At one level, this means that without the energy from the Sun, the Earth would soon run down. Eventually, the Sun and, possibly, the universe, will exhaust its resources and fizzle. Or, put another way, no matter how much you straighten your room this week, you're going to have to straighten it again next week.

KINETIC THEORY OF GASES

The caloric theory of heat finally met its end around 1860, when James Clerk Maxwell (1831–79) and Ludwig Boltzmann (1844–1906) independently used a series of equations to describe the behavior of gases more completely than anyone ever had before. The temperature of a gas, Maxwell said, does not reflect the speed of movement of all the molecules of a gas uniformly. Instead, it reflects the statistical average of their movement, in all directions and at all velocities. When a gas is heated, he explained, the molecules move faster and bump into each other more, and this bumping increases the pressure of the gas.

MAXWELL'S DEMON

In 1871, Maxwell invented a tiny character—which has come to be known as Maxwell's demon—to illustrate the statistical nature of both entropy and his kinetic theory of heat in gases. Imagine a gas evenly distributed in both chambers of a two-room house. Only one opening exists, and that is a sliding door between the two rooms. As described in Maxwell's kinetic theory, some of the gas molecules in both rooms are moving slowly, while others are moving rapidly. As the molecules float (or speed) by, the demon grabs the slow ones and puts them in the other room. From the other room he grabs the fast molecules and pulls them through the opening into the first room. In this way, eventually, one room is full of cold (slow-moving) molecules

GREAT MOMENTS IN THERMODYNAMICS

1822	Jean-Baptiste Joseph Fourier publishes equations of heat flow
1824	Nicolas Léonard Sadi Carnot's theorem becomes the basis for the independent formulations of the second law of thermodynamics by Clausius and Kelvin
1847	James Prescott Joule experimentally establishes the mechanical theory of heat (the "mechanical equivalent of heat")
1847	Hermann von Helmholtz outlines first law of thermodynamics (the law of conservation of energy)
1850–51	Rudolf Clausius and William Thomson (Lord Kelvin) formulate the second law of thermodynamics
c. 1860–70	James Clerk Maxwell and Ludwig Boltzmann independently develop the kinetic theory of gases
1871	Maxwell's demon is presented in his *Theory of Heat*

and the other is full of hot (fast-moving) molecules. If such a demon existed (which, of course, it does not) you could heat a room without using any energy.

From the caloric theory of the 18th century, physics had come a long way in a little over 70 years of studying the nature of heat and its interrelations with other forms of energy. Based on the power of the atomic theory and through the use of mathematics, models and careful experimentation, two abiding principles had been achieved that provided substantially greater insight into the workings of thermodynamics.

MAGNETISM, ELECTRICITY AND LIGHT

All of Europe was experimenting with electrical current in 1819, when Hans Christian Ørsted began teaching his physics class at the University of Copenhagen. And Ørsted was no exception. In a classroom demonstration, he took a wire through which he ran an electric current and brought it close to a compass needle. Speculation had been in the air for a long time that some relationship existed between electricity and magnetism. And Ørsted probably suspected the electrical current and the magnet would have some effect, one on the other. He was right.

With a sudden and immediate reaction the compass needle swung, not in the direction of the current, which streamed steadily from Ørsted's voltaic cell, but instead into a position at a right angle to the current. Ørsted reversed the direction of the current. The compass needle swung again, this time in the opposite direction, again at a right angle.

Ørsted had demonstrated for the first time, before his students, that a connection exists between electricity and magnetism, and he opened the door to a new study: electromagnetism. It would prove to be the most productive area of study in the 19th century.

AN ANCIENT MYSTERY

The study of both electricity and magnetism dated as far back as the work of William Gilbert of Colchester (1544–1603) in the 16th century. While the ancient Greeks had known about the magnetic properties of amber, Gilbert showed that many other substances could be magnetized—sulfur, glass, iron—and he was the first to use the terms *electric force*, *electric attraction*

In William Gilbert's De Magnete, *published in 1600, he explored the nature of magnetism. In this illustration from his book, a blacksmith magnetizes a glowing iron bar by pounding it with the ends pointing north* (septentrio) *and south* (auster). (Courtesy of the Burndy Library)

and *magnetic pole*. Often thought of as the founder of the study of electricity, he wrote about his researches in his book *De Magnete*, published in 1600.

In the 17th century Otto von Guericke devised a machine that could generate static electricity, and in 1745 both Pieter van Musschenbroek (1692–1761) and Ewald von Kleist (1700–48) independently discovered the principle of the Leyden jar, in which static electrical charges could be stored. With the Leyden jar, scientific (as well as popular) interest in electricity swelled, and Benjamin Franklin (1706–90) did extensive research exploring the nature of its positive and negative polarity, its relationship with magnetism, its ability to melt metals and so on. He also demonstrated, with his famous kite experiment, that lightning is electricity.

Not until Volta's invention of the voltaic cell (see Chapter One), though, did it become possible to create a continual, steady source of electricity; up to that point all sources of electricity were static. Before Volta, electricity could be stored, but it could only be discharged in a single (often mighty) jolt.

The big breakthroughs with electricity, however, still lay ahead in the 19th century. Not only would electricity, once harnessed, change the way people lived, but from a new understanding of electricity, magnetism and their relationship would emerge powerful new theories that would transform the way people thought about the universe. The first giant steps in this direction were taken by a young man named Michael Faraday.

FARADAY, THE GREAT EXPERIMENTER

One of 10 children, the son of a blacksmith in England, Michael Faraday (1791–1867) started life with no hope of going to school beyond learning to read and write, much less obtaining a university education. At the age of 12, he'd begun earning his share of the rent, and his school days were finished. But some people have such curious minds that nothing can keep them from trying to find out what the world is made of, or why people act

Michael Faraday
(AIP Emilio Segrè Visual
Archives, E. Scott Barr
Collection)

the way they do, or what makes things work. And Michael Faraday had one of those tirelessly curious minds. He also had a bit of luck: He found a job as an apprentice to a bookbinder, and as he bound the outside of the books, he avidly devoured the words inside. He read the articles on electricity in the *Encyclopaedia Britannica* and Lavoisier's *Elements of Chemistry*. He also read (and bound) a book called *Conversations on Chemistry* by Mrs. Marcet, a Swiss doctor's wife, whose popularization of science was widely read in 1809, when it was published.

Then another piece of luck came Faraday's way. A customer of the bookbinder gave young Faraday tickets to four lectures by Humphry Davy at the Royal Institution. Faraday was elated and took scrupulous notes at all four lectures, which he later bound and sent to Davy, with an application for a position as assistant at the institution. A few months later, when an opening came up, Davy offered the job to Faraday. "Let him wash bottles," one of Davy's colleagues said. "If he is any good, he will accept the work, if he refuses, he is not good for anything." The job paid less than his bookbinding job, but Faraday jumped at the chance.

Shortly thereafter, in 1813, Davy set off for Europe with Faraday at his side as secretary and scientific assistant. Though Davy's wife treated Faraday as a servant, the young man never complained, instead taking advantage of the opportunity to meet the key figures of science, including Volta, Ampère, Gay-Lussac, Arago, Humboldt and Cuvier. As they traveled from laboratory to laboratory across Europe, performing experiments and attending lectures, Faraday received the education he had never had.

On their return, in 1815, Faraday became, officially, assistant in the laboratory and mineral collection and superintendent of the apparatus at the Royal Institution. He was Davy's right hand in the laboratory, adroit, expert and dedicated, often working from nine o'clock in the morning until eleven at night. After a few months he received a raise to £100 a year, and he remained at that salary until 1853.

When Faraday read of Ørsted's experiment in 1820, he—like all the rest of the scientific community—became very excited. Ørsted's compass showed that electric current was not traversing the wire end to end in a straight line, as everyone had supposed, but instead was circling the wire. Further demonstrations by André Marie Ampère [ahm PAIR] in Paris corroborated this idea. Ampère showed that if two electrical wires are strung parallel, with one loose enough to move freely, when the current in both wires runs in the same direction, the wires are drawn together; when the current runs in opposite directions, the two wires are pushed apart.

Faraday set up a simple experiment of his own. In September 1821 he demonstrated "electromagnetic rotation," showing that a wire could be made to move around a fixed magnet through the use of electric current,

Faraday in his laboratory at the Royal Institution   (AIP Emilio Segrè Visual Archives)

and that a magnet could be made to move around a fixed wire. It was the first primitive electric motor.

Unfortunately, Davy became angry with Faraday over this experiment, claiming that Faraday had overheard a discussion between Davy and William Wollaston describing a similar experiment. Faraday admitted he may have gotten a start from the conversation, but his apparatus was substantially different, and both Wollaston and history seem to agree on this point.

In any case, it is perhaps the least of Faraday's discoveries. He was stalking a much bigger conquest. In 1822 Faraday wrote in his notebook: "Convert magnetism into electricity." Ørsted had used electricity to create magnetism (the compass needle responds to magnetic force); couldn't the reverse process also take place? Starting from ideas already set forth in part by Ampère and another physicist, William Sturgeon, Faraday began with an iron ring, wrapping one segment of it with a coil of wire. He could introduce an electric current into the wire by closing a circuit with a key. He then wrapped another segment of the ring with wire and connected it to a galvanometer. He thought that the current in the first coil of wire might cause a current in the second coil. The galvanometer would measure the presence of the second current and tell the story.

This idea did work—it was the first transformer—but the results contained a surprise. Despite the steady magnetic force set up in the iron ring, no steady electric current ran through the second coil. Instead, a flash of current ran through the second coil when Faraday closed the circuit—with a jump on the galvanometer. Then when he opened the circuit again, another flash of current, marked by a second galvanometer jump.

Since Faraday knew no mathematics (having never gone to school), he used visualization to explain this phenomenon—and came up with the idea of lines of magnetic force. He had noticed that if you sprinkled a paper with iron filings, held it over a strong magnet and tapped it, the filings would arrange themselves in distinct patterns, along what Faraday concluded were the magnet's lines of force. He conceived of the idea that an electric current forms a kind of magnetic field radiating out in all directions from its source. When he closed the circuit in his experiment, lines of force radiated out, but the second coil of wire cut across them. When that happened, a current was induced in the second coil. When he opened the circuit, the lines of force "collapsed back," and again, the second wire cut through their path and a current was induced. He saw how he could work out what the lines of force would look like for a bar magnet, for a spherical magnet such as the Earth, for an electric wire. And for the first time since Galileo and Newton had conceived of the mechanistic universe, now a new and even more productive way to look at the universe—field theory—was in the making.

During one of the many enormously popular lectures he was now giving at the Royal Institution, in 1831, he demonstrated the lines of force in another way. He took a coil of wire and moved a magnet into the coil. The needle of the galvanometer attached to the wire swung, then stopped when the movement of the magnet stopped. When he moved the magnet out of the coil, again the galvanometer registered. Moving the magnet around inside the coil registered. If he moved the coil of wire over the magnet, presence of a current showed up. But if he let the magnet just sit motionless inside the coil of wire, no action registered on the galvanometer; there was no current. Faraday had discovered the principle of electromagnetic induction. That is, he had found that by combining mechanical motion with magnetism he could produce electric current. This was the basic principle of the electric generator or dynamo. (Another physicist, Joseph Henry, across the ocean in the United States, had also come up with an excellent demonstration of this same idea but had set it aside without publishing. As a result, Faraday, who never lost focus and pursued his work with extraordinary single-mindedness, gets the credit, which Henry freely acknowledged.)

Faraday's next step, of course, was to build a generator, producing a continuous source of electricity instead of the jerky, on-off variety he had induced in his experiment. This he did by setting up a copper wheel so that its edge passed between the poles of a permanent magnet. As long as the wheel turned, an electric current was set up in it, and the current could be led off and set to work. By adding a water wheel or steam engine to turn the wheel, the kinetic energy of falling water or the combustive energy of burning fuel could be transformed to electrical power. Electrical generators today don't look much like Faraday's original model, and it took some 50 years for practical applications to be found, but it was unquestionably the most important electrical discovery ever made.

From childhood, Faraday had a profound belief in the interconnection and unity of natural forces and phenomena, and he recognized that his field theory, which he first published in 1844, and his explorations of the interrelatedness of magnetism, electricity and motion contributed to this vision. In the opening paragraph of his paper "On the Magnetization of Light and the Illumination of Magnetic Lines of Force," read before the Royal Society on November 5, 1845, he wrote:

> *I have long held an opinion, almost amounting to conviction, in common I believe with many other lovers of natural knowledge, that the various forms under which the forces of matter are made manifest have one common origin; or, in other words, are so directly related and mutually dependent, that they are convertible, as it were, one into another, and possess equivalents of power in their action.*

At first, not many people took Faraday's field theory seriously, but in many ways, Faraday's belief in the fundamental unity of nature was vindicated by the work of Joule, Thomson, Helmholtz, Clausius and Maxwell in the following decades.

Meanwhile, the relationship between Faraday and Davy had continued to deteriorate. As time passed, Davy must have recognized that Faraday was passing him up, and he began to become jealous and bitter. When Faraday's name came up for admission as a fellow to the Royal Society, Davy opposed; but Faraday was made a fellow in 1824, despite Davy's lone dissenting vote. In 1825 Faraday became director of the laboratory, and he became professor of chemistry at the Royal Institution in 1833. A gentle and religious man, who preferred to spend time in his laboratory or at home in the companionship of his wife, Sarah Barnard, Faraday never responded to Davy's shabby behavior. He had other things to do. As his successor at the Royal Institution, John Tyndall, described him, Faraday "was a man of excitable and fiery nature; but through high self-discipline he had converted the fire into a central glow and motive power of life, instead of permitting it to waste itself in useless passion."

To Faraday, the great experimentalist, we owe a giant debt. As Ernest Rutherford said in 1931:

> *The more we study the work of Faraday with the perspective of time, the more we are impressed by his unrivalled genius as an experimenter and a natural philosopher. When we consider the magnitude and extent of his discoveries and their influence on the progress of science and industry, there is no honour too great to pay to the memory of Michael Faraday—one of the greatest discoverers of all time.*

THOMAS ALVA EDISON (1847–1931)

Thomas Edison wasn't very interested in science. And he didn't care much about the nature of electricity—that was for the scientists to figure out. What he wanted to do was tame it and make it jump through hoops. He wanted to put it to work. Work could have been Edison's middle name. The child of poor parents, Edison, born in Ohio, was taken out of school by his mother and at age 12 found himself a job working as a newsboy selling papers on a train in Michigan. Not content simply to sell papers, he started publishing his own on board the train as it made its passenger run between Port Huron and Detroit. Newspapers weren't Edison's pri-

Thomas Alva Edison (left) in his lab. (With him is German-American electrical engineer Charles Steinmetz.) (General Electric Research Laboratory [Courtesy AIP Emilio Segrè Visual Archives])

THE SCOTTISH THEORIST

James Clerk [pronounced CLARK] Maxwell was born in 1831, the year Michael Faraday made his most influential discovery, electromagnetic in-mary interest, though, and he put his profits into buying chemicals and setting up a small laboratory in the train's baggage car. His chemical work was short lived when one of his experiments resulted in an explosion that nearly wrecked the baggage car, and both the young experimenter and his chemicals were hastily kicked off at the next stop.

By 1862 his interest had shifted to the new field of telegraphy and his reputation as the fastest and most accurate telegrapher in the country earned him enough money to begin collecting books on electricity, including the collected works of Faraday. His next venture found him in New York in 1869, where he offered his first major invention, an improved stock-ticker, to a big Wall Street firm. He wanted $5,000 for it, but before he could ask his price the president of the firm told him he couldn't offer more than $40,000! Edison took it. By the time he was 23 he was solidly in the inventing business with his own small firm of consulting engineers.

Edison usually put in 20 hours a day at his work and by 1876 had expanded his interests to set up a research laboratory in Menlo Park, New Jersey. And there his true genius flowered. Out of his so-called invention factory, the world's first full-scale private research facility, a steady stream of new inventions soon flowed. Working with a staff of engineers (numbering more than 80 at its peak), the "Wizard of Menlo Park" patented more than 1,300 inventions before he died. Not always the most likable of men, brusque to friends and ruthless to his competitors, he nevertheless was responsible for an amazing number of electric inventions and products that changed the life-style of the world—including the phonograph in 1877 and the incandescent electric lamp in 1879. The following year he illuminated the main street of Menlo Park with electric lights to the astonishment of reporters from around the world. And, in 1881, at a location on Pearl Street, New York, he built the world's first central electrical power station. In the 1890s Edison began making America's first commercial motion pictures using his Kinetoscope process, and he began showing the films, which could only be viewed in a small viewing machine by one person at a time, in a "Kinetoscope Parlor" in New York in 1894.

In 1960 Edison was elected to the Hall of Fame for Great Americans, a tribute not only to his major inventions but also to the hundreds of others, large and small, that, in the words of the United States Congress, forever "revolutionized civilization."

duction. As a child Maxwell was so brilliant in mathematics that he didn't seem to have good sense: His classmates called him Daffy. At 15 he submitted to the Royal Society of Edinburgh a paper on the drawing of oval curves that was so impressive that many members of the society thought it couldn't have been written by someone so young. By the time Maxwell was in his early thirties, he had already explained the probable nature of Saturn's rings (1857), about which he was right, and had developed, independently of Ludwig Boltzmann, the kinetic theory of gases (1859–60).

But he had always been intrigued by Faraday's work; in December 1855 and February 1856, as a 24-year-old fellow at Trinity College, Cambridge, Maxwell had presented an extraordinary paper entitled "Faraday's Lines of Force." And now, between 1864 and 1873, Maxwell would bring his mathematical wizardry to bear on Faraday's speculations about electromagnetic lines of force, providing the theoretical justification it needed.

In the process, Maxwell worked out a series of simple equations that described all the observations made about both magnetism and electricity and illustrated that the two forces were inextricably bound together. This monumental work, known as the electromagnetic theory, showed that magnetism and electricity could not exist separately.

In support of Faraday's field theory, Maxwell showed that an electromagnetic field was, in fact, created by the oscillation of an electrical current.

James Clerk Maxwell, whose electromagnetic theory transformed the study of physics (AIP Emilio Segrè Visual Archives)

This field, he added, radiated outward from its source at a constant speed. This constant speed could be calculated by taking a ratio of certain magnetic units to certain electrical units, which worked out to be approximately 186,300 miles per second. Light travels at 186,282 miles per second—a coincidence too astounding, Maxwell thought, to be accidental. From this he came to the conclusion that light itself must be an oscillating electric charge. Light, he concluded, was electromagnetic radiation! This point he was unable to prove, but it seemed a strong hunch—and his hunch was proved a generation later.

But Maxwell went farther. Light, he postulated, was probably just one of a large family of radiations caused by charges oscillating at different velocities. (Evidence had already been found that there might be a lot more than we could see: William Herschel had discovered infrared light, invisible to the naked eye, in 1800; Johann Ritter had found ultraviolet light, also invisible, at the other end of the spectrum in 1801.)

Maxwell published his *Treatise on Electricity and Magnetism*, on his theory of electromagnetism, in 1873. It was a brilliant work, adding the precision of mathematics and quantitative prediction to Faraday's views on field theory and, specifically, on electromagnetism. Like Thomas Young, he postulated the existence of ether throughout space as the medium through which electromagnetic waves traveled; this postulation was later disproved, but his equations don't depend on ether's existence, and they hold up as well as ever in the everyday world of "classical" physics (though not in Einstein's physics of relativity or the world of quantum mechanics).

In one of those strange coincidences of history, James Clerk Maxwell died in 1879, the same year that another great theoretician, Albert Einstein, was born. Like Maxwell's work in the 19th century, Einstein's would dominate much of the century to come. Maxwell didn't live long enough to see his theories validated experimentally (he died of cancer before he was 50), but the proof was not far off. Less than a decade later a young physicist in a laboratory in Germany did the job.

HERTZ'S WAVES

A student of Helmholtz, Heinrich Rudolf Hertz (1857–94) became interested in Maxwell's equations concerning electromagnetic fields in 1883. Helmholtz suggested that Hertz try for a prize offered by the Berlin Academy of Science for work in electromagnetics, and Hertz, by now teaching at Karlsruhe, decided to give it a try. In 1888 he found something. Hertz set up an experiment that made it possible for him to detect the presence of the type of long-wave radiation that would be produced if light

YOUNG, FRESNEL AND LIGHT WAVES

Light, nearly everyone knew, was made of particles; Newton had established that long ago. For one thing, light could not pass around corners, as sound waves did, and it cast sharp shadows. So when Thomas Young (1773–1823) began to think that light might be a wave instead, he had a sharp uphill battle to fight.

The wave versus particle controversy was age-old, though. (Francesco Grimaldi, 1618–63, had observed that a beam of light passed through two narrow apertures became slightly wider than the apertures, indicating a slight bend, which he called "diffraction.") And some people believed the jury was still out on the issue.

As a child, Thomas Young was a remarkable prodigy who began reading at age two, had read the Bible twice by the age of six and learned a dozen languages, including Persian and Swahili. In later life, this tremendous facility with language served him well when in 1814 he tackled deciphering the hieroglyphics of the ancient Rosetta Stone found by Napoleon's expedition to Egypt in 1799.

Early in his career Young lectured from 1801 to 1803, with Davy, at Rumford's Royal Institution, and during these years he investigated the anatomy of the eye (finding that astigmatism is caused by imperfections in the cornea), color theory (founding, with Helmholtz, the Young-Helmholtz three-color theory that later became the basis for color television and color photography), and the nature of light.

To test the particle versus wave question, Young performed a test, sometimes known as his "fringe experiment," in which he shined a light through several narrow openings. At the edge of these openings, several blurred bands of light appeared. If light were particles and not waves, only a clear, sharp shadow should have appeared. Particle theory had no explanation for this new demonstration of light diffraction. Then Young went further. He thought about the way two pitches of sound sometimes cancel each other out. (An especially good example is the way two sounds on a public amplification system may start out together, screeching loudly,

were in fact a type of electromagnetic radiation. He also devised a way to measure the shape of the wave if it appeared.

It did appear and he was able to measure it. The wavelength was 2.2 feet (66 centimeters)—one million times the size of a wavelength of visible light. Hertz was able to show, as well, that the waves he measured involved both electric and magnetic fields and therefore had an electromagnetic nature.

and then suddenly produce nothing but silence.) The reason for this is that two different pitches of sound travel in waves of different length. They may start together with the peaks of the waves matched, then get out of sync so that the trough of one coincides with the peak of the other, and they cancel each other, producing no sound. Young thought that if light were produced by waves as well, the same kind of interference might occur. Young projected light beams through two narrow openings and shined them on a wall. The two beams overlapped, and where they overlapped, stripes of light and dark appeared, indicating that interference existed for light exactly as it did for sound. Thomas Young had resurrected the wave theory of light.

Young also suggested that light traveled in transverse waves—that is, at right angles to the point from which it emanated, much the way waves of water do in the ocean—not in longitudinal waves as sound does. This idea helped resolve some questions about polarized light and double refraction, and it also had implications, once Maxwell's theory of electromagnetism was established, for the entire electromagnetic spectrum, of which light is only a part.

Initially, especially in Newton's native England, most scholars laughed at Young's wave theory. It remained for Augustin Fresnel, in France, to provide the needed mathematical backup. But light wave theory still had problems. If light consisted of waves, not particles, then the waves had to travel in some medium, such as air provides for sound waves and water provides for waves in the ocean. (There is no air outside the atmosphere to transmit light from stars.) Early wave theorists—including Maxwell much later in the century—thought there must be an "ether" filling all space, through which light undulated, but no proof of that ether's existence had ever been found. Meanwhile, though more and more scientists turned to the wave theory, light still seemed to behave sometimes as if composed of particles. Albert Michelson and Edward Morley would show in 1887 that the suspected ether did not exist, and the continuing question of light's nature led to some of the most profound discoveries of the next century.

What Hertz had found were not light waves, but, as it turned out, radio waves, for which Marchese Guglielmo Marconi found a use in 1894 as a kind of wireless communication. (*Radio* is short for radiotelegraphy—that is, telegraphy by radiation rather than telegraphy by electric currents.)

Hertz had succeeded in proving the existence of electromagnetic waves, verifying the validity of Maxwell's equations. And another piece of the great puzzle of physics slipped into place.

Heinrich Hertz
succeeded in proving
the existence of
electromagnetic waves,
verifying the validity of
Maxwell's equations.
(Deutsches Museum AIP
Niels Bohr Library)

Throughout the 19th century a new pattern had begun to emerge in physics, a pattern of an idea set forth, validated by experiment and reinforced by mathematical theory. It was a threefold process that increasingly gained the respect of the scientific community, and it worked for Mayer and Joule's heat equivalent; for Faraday, Maxwell and Hertz's work on electromagnetism; and for Young and Fresnel's insights into the nature of light.

The most stunning accomplishment of the century was the steady untangling of threads by many hands—and the extraordinary genius of Faraday and Maxwell—that led to recognition of that great underlying force, electromagnetism. Nearly every phase of our lives is touched by Faraday's electric motor, transformer and generator. But the significance of the underlying concepts of field theory and electromagnetism count among the most telling insights in the history of humankind's investigation into the nature of the universe.

C H A P T E R   5

SKY AND
EARTH

For as far back as we have records, people have been watching the skies, trying to understand the specks of light they saw above the night horizon. By the 19th century, theory had come a long way, with Copernicus's recognition, published in 1543, that the Sun, not the Earth, resided at the center of the Solar System; Kepler's work on orbits, published in 1609; and Kant's 18th-century work on nebulas.

The telescope, first used for astronomy by Galileo in 1616, gave a giant boost to astronomy. Now the four great moons of Jupiter could be seen, as could the rings of Saturn and the surface of the Moon. William Herschel's surprising discovery in 1781 of a seventh planet, Uranus, was made thanks to the improved telescope he used.

Uranus's orbit was strange, however. It seemed to indicate that at least one more planet existed in the Solar System. But where was it?

And what were the cloudlike features catalogued so extensively by Charles Messier in the 18th century? Were they vast star systems so far away that they appeared in the telescope as only a blur? Or were they just clouds of gas, as some supposed? How could anyone tell?

What was the Sun made of? And the stars?

Better detection methods lay at the heart of further progress. Greater precision was called for, as were improved calculations and better instruments. To the challenge of these demands rose numerous dedicated, passionate and incisive minds. But in the 19th century, two other unusual advances boosted results dramatically for astronomers: a way, surprisingly enough, to determine what elements the stars were made of (through spectroscopy) and a way to record what their telescopes pointed at (through photography, invented in 1826).

MARIA MITCHELL: FIRST WOMAN ASTRONOMER IN THE UNITED STATES

Maria Mitchell was born in Nantucket, Massachusetts in 1818. She had little opportunity for a formal education, but she was lucky enough to have a father who taught her, and she became the librarian at the Nantucket Atheneum. But she also learned to love watching the skies and made many amateur observations.

Then, on October 1, 1847, Maria Mitchell discovered a comet. She was 29. Immediately she gained the attention of the scientific community, and in 1849 she got a job at the U.S. Nautical Almanac Office. There she made astronomical computations and earned a reputation for her competence and accuracy. In 1865 she was named professor of astronomy at the newly formed Vassar College.

Maria Mitchell's achievements in astronomy may have been moderate, but her early contribution to the idea of women as professional scientists was a landmark. She succeeded in cutting through the prejudice of her time by doing what she loved, despite society's expectations that women should stay home, keep house and raise children. Instead, she was the first woman admitted to the prestigious American Academy of Arts and Sciences, and until her death in 1889, she continued to teach other women at Vassar that science is for everyone.

SEEING BETTER

An extraordinary proportion of advances in astronomy in the 19th century can be traced to the optical shop of one dedicated "lens grinder," a man whose name comes up whenever chemistry or physics or astronomy of this era is discussed: Joseph von Fraunhofer. Not only, as described earlier, did this once-penniless orphan boy grow up to discover the spectral lines that bear his name, but he was known far and wide for his precision-ground lenses and supremely crafted telescopes encased in red Moroccan leather.

Using one of Fraunhofer's telescopes, Friedrich Bessel (1784–1846) made the first successful measurement of distance to a star, known as star number 61 in the constellation Cygnus (61 Cygni). For three centuries astronomers had been trying to determine the parallax of a star—any star. By finding the parallax (an object's apparent shift in position viewed from two different locations), an astronomer could make use of triangulation to determine the star's distance from the Earth. But stars were so far away

Maria Mitchell was the first professional woman astronomer in the United States.
(Photo courtesy of Virginia Barney; print courtesy H. Wright)

that, even measuring from six months apart on the Earth's orbit—which is the largest baseline available to earthbound astronomers—no good measurement had ever been made. Bessel chose 61 Cygni, even though it is a relatively dim star, because of its rapid proper motion, the most rapid of any star. He trained his trusty Fraunhofer on it, using a special instrument called a heliometer that he had designed, but Fraunhofer had made. Through painstaking observations, Bessel measured the tiny displacements made by 61 Cygni and was able to compare its position with two other, even fainter, stars close by. To his amazement, 61 Cygni's parallax indicated that it was six light years away (or six times the distance light can travel in a year—six times a little less than 6 trillion miles). Since Kepler had estimated that the stars were about a tenth of a light year away and Newton had thought the distance was about two light years, this discovery began to change astronomers' ideas drastically about the size of the universe.

When Bessel announced his achievement in 1838, one more piece of the Copernican puzzle slipped into place, since even a small parallax for a star indicated that the Earth was moving through space.

Also with his heliometer, Bessel noticed that two stars, Sirius and Procyon, each had small deviations that couldn't be accounted for as parallactic but were more like wobbles. In 1841 Bessel postulated that each of these two stars was revolving around an unseen companion.

The rest of that story belongs to a second precision lens maker, Alvan Clark (1832–97) of Massachusetts, who, like Fraunhofer, made world-renowned lenses. One night in 1862 Clark was testing an 18-inch lens he and his father were working on, when he trained it on Sirius and spotted a tiny speck of light nearby. It was the companion star whose existence Bessel had suggested 21 years before.

Two more major discoveries were made with Clark telescopes. In 1877, during a close approach of Mars, Asaph Hall (1829–1907) of Connecticut, at the insistence of his wife, Angelina Stickney (his former mathematics

Edward Emerson Barnard (©UC Regents; Lick Observatory image)

teacher) to "try it just one more night," discovered the two satellites (moons) of Mars. And in 1892 Edward Emerson Barnard (1827–1923) discovered a fifth moon of Jupiter, the first to be discovered in three centuries.

William Parsons, the third earl of Rosse, was another matter. His great love was building giant reflector telescopes by himself, beginning in 1827. By 1845 he had succeeded in building a true giant, a 72-inch reflector he called "Leviathan." Despite the constant foggy weather in his native Ireland, in 1848 Lord Rosse was able to study the Crab Nebula, which he named, and he identified several spiral-shaped objects that turned out to be very far-off galaxies.

Fraunhofer and Clark's success with improved lenses inspired the construction of giant refractor telescopes at the end of the century, including the Lick Observatory in California, built in 1888, with a 36-inch aperture, and, in 1895, the Yerkes Observatory near Chicago, the construction of which Clark supervised. The Yerkes Observatory has the largest refractor telescope ever built, with a 40-inch aperture.

MISSING PLANETS

When they looked at the skies, the ancients saw certain objects they called "wanderers," or planets, that traveled in a strange manner across the canopy above, and they named them Mercury, Venus, Mars, Jupiter and Saturn. Of the planets we know today, of course, they also knew Earth, although no one then thought of *terra firma* as a planet. William Herschel astonished everyone by discovering a seventh planet, Uranus, in 1781. (He wasn't the first to see it, actually. It can be seen without binoculars or a telescope. But he was the first to identify that it was a planet.) Herschel had used systematic searching, a good telescope, excellent eyesight and the help of his sister, Caroline.

But perhaps there were more planets. Many astronomers were disturbed by an aberration in the orbit of Mercury, and Urbain Jean Joseph Le Verrier (1811–77) became sure that the presence of another planet between Mercury and the Sun would account for it. He made calculations and predicted its orbit and its size (1,000 meters in diameter) and named it Vulcan. But try as they might, no one ever found it. (Einstein would later explain why Mercury's orbit did not fit with Newtonian physics.)

Uranus's orbit had much the same problem, and with this Le Verrier had much better luck. Again, he ran mathematical calculations and drew up equations. Then he contacted Johann Galle (1812–1910) in Berlin and told him where to look, and on September 23, 1846, they were in luck. Almost exactly where Le Verrier had said to look, Galle found the planet Neptune, another gaseous giant of nearly the same size as Uranus. (At Cambridge, John Couch Adams had made the same calculations a few months before

Joseph Le Verrier, the discoverer of Neptune, the eighth planet
(Courtesy of Yerkes Observatory)

but had not had access to a telescope to test out his thesis.) It was a triumph for astronomy as a science.

FRAUNHOFER'S LINES

When Fraunhofer died at 39 on June 7, 1826, he left behind a legacy not only of his exquisite lenses, but also his mysterious lines. Then in 1859, Gustav Kirchhoff and Robert Bunsen announced the invention of their spectroscope, opening up the discovery of numerous elements.

One evening Kirchhoff and Bunsen had been working in their laboratory in Heidelberg when they saw a fire burning in the nearby city of Mannheim, 10 miles away. They trained their spectroscope on it and discovered that they could detect the presence of barium and strontium from the arrangement of the fire's spectral lines—even at such a distance. Would it be possible, Bunsen began to wonder, to train the spectroscope on the light of the Sun and detect what elements were present there? "But people would think we were mad to dream of such a thing," he muttered.

In 1861 Kirchhoff put the idea to the test, and in the light from the Sun he succeeded in identifying nine elements: sodium, calcium, magnesium, iron, chromium, nickel, barium, copper and zinc. Amazingly enough, the great source of light in the sky, once worshipped by ancient peoples as a god, contained many of the very same elements as the Earth. Gustav Kirchhoff had opened the door to two new sciences, spectroscopy and astrophysics, and had established another link between earthly physics and chemistry and the stars. It was another stunning example of convergence of worlds that once had seemed entirely separate.

An English amateur astronomer named Sir William Huggins first put the spectroscope to use on deep-sky objects in 1864. A wealthy man, he had a private observatory, complete with telescope, located on a hill in London. He fitted a spectroscope to his telescope and studied the spectral lines emitted by two stars so bright they can be easily seen with the naked eye, Aldebaran and Betelgeuse. He could identify the thumbprint of the elements

William Huggins
(©UC Regents; Lick
Observatory image)

THE BRILLIANT MARY SOMERVILLE

It wasn't easy for women to open the doors of science in the 19th century. Mary Somerville (née Fairfax) (1780–1872) not only opened them for herself but became the favorite science writer of such scientific greats as Charles Lyell and John Herschel. The daughter of a Scottish admiral, Mary had no schooling until she was 10 and couldn't even read until she was 11, but she hadn't wasted even those youthful years. She collected fossils and stones and somehow got hold of a celestial globe and began to study astronomy.

As soon as she learned to read, doors opened even wider as she taught herself Latin and Greek. Taking time also to learn to play the piano, she approached it with more than traditional drawing room art, even learning to tune it and repair broken strings herself.

Her earliest real love, though, was mathematics. She discovered algebra and geometry on her own and quickly mastered the writings of Euclid. Needless to say all this intellectual and artistic activity was a little disconcerting to those around her—enough so that she was talked into marrying a rather stuffy and traditional friend of her father, Samuel Greig. With the precocious Mary now someone else's problem, her family started to rest easy again. About how Greig managed not much is known, but he died when Mary was only 33, leaving her a very wealthy widow. No lavish balls and fancy dress parties for the young widow, though; the money had a much better use. In no time at all she had purchased enough books to build a wonderful mathematics library.

Marrying a second time, and by her own choice, she had better luck. William Somerville was an army doctor and a scholar; he also respected his wife's intelligence and encouraged her mathematical and scientific pursuits.

---

iron, sodium, calcium, magnesium and bismuth. Next he tried a nebula and, with a feeling of great suspense and awe, stepped up to look. "Was I not about to look into a secret place of creation?" he wrote in his journal. Perhaps this moment would settle once and for all the question of which theory about nebulas was correct.

> *I looked into the spectroscope. No spectrum such as I expected! A single bright line only! . . . The riddle of the nebulae was solved. The answer, which had come to us in the light itself, read: Not an aggregation of stars, but a luminous gas. Stars after the order of our own sun, and of the brighter stars, would give a different spectrum; the light of this nebula had clearly been emitted by a luminous gas.*

In 1816 when William was transferred to London, Mary suddenly found herself in the center of English scientific life, and she knew what she was going to do. She would write about science. She published *The Connection of the Physical Sciences* in 1834. Her *Physical Geography* (1848) quickly won many scientific admirers, although it was attacked by some of the clergy who were still fighting the old battle against an ancient age for the Earth. In fact the book did so well that an organization called The Society for the Diffusion of Useful Knowledge asked her to write a book on astronomy for them. It, too, won admirers, and she followed with a book on Newton's *Principia*. During the 1820s a woman could not become a professor, but Mary Somerville quickly became a favorite writer of scientists and professors. With the growing specialization of science, it was becoming difficult for the curious scientist to know what was going on in other fields. Drawn by the care and precision of her writing, her brilliant understanding of the facts and her lucid explanations, more than a few scientists turned to Mary Somerville. John Herschel, excited by her manuscript *Mechanics of the Heavens* (1831), became a good friend and ardent supporter, recommending not only that book but others of hers to his friends. Soon, not just scientists were reading Mary Somerville, but also the literate public who wanted clear, factual and readable reports on the growing scientific explosion around them.

A science writer of the highest intelligence and wit, Mary Somerville soon found a place at the heart of London's scientific society as the Somerville home became a gathering place for some of the finest minds of her time. Unable to attend a university, Mary Somerville now had a good part of the university attending her—to the mutual delight and benefit of all.

Huggins unfortunately got off on the wrong foot by assuming that, because this nebula was gaseous, all nebulas—including those with elliptical and spiral shapes—were made of gas. But, nonetheless, the first use of spectroscopy in astronomy was a stunning success. Functioning for astronomical research in much the same role as fossils do in geology, Fraunhofer's lines and the spectroscope provide invaluable information on temperatures, compositions and motions of gaseous nebulas and stars. As Kirchhoff showed, a hot, glowing, opaque object emits a continuous spectrum—a complete rainbow of colors without spectral lines. Viewed through a cool gas, though, dark spectral absorption lines appear in the spectrum. These reveal the chemical makeup of the gas. If, however, the gas is viewed at an angle, a different set of patterns appear. These tools have become a sort of Rosetta stone for astronomers studying the gaseous nebulas.

PHOTOGRAPHING THE STARS

John Herschel (1792–1871), son of William Herschel, was one of the first to recognize the possibilities photography had for astronomy. Although photography was discovered in 1826, it wasn't used for astronomy until the 1840s. Its great advantage was, of course, that working from photographic plates, astronomers no longer had to work in real time. They could pore over the plates for hours, using a magnifying glass or microscope, and note differences in plates taken at different times. As the medium became more versatile, it became possible to leave a plate exposed for periods of time to catch objects that could not otherwise be seen, even with a telescope. In 1895, Edward Emerson Barnard photographed the Milky Way for the first time. Photography would become more and more important as a tool for astronomers in years to come.

SECOND-GUESSING THE SUN

But the closest star, of course, is our own Sun, and the 19th century saw two discoveries that added to the understanding of solar physics.

John Herschel
(©UC Regents; Lick Observatory image [Mary Lea Shane Archives of Lick Observatory])

In 1843, Samuel Heinrich Schwabe announced his discovery of the cyclic action of sunspots. The discovery marked the beginning of early work in solar physics and astrophysics. And in 1868, Pierre Jules César Janssen discovered helium while studying the spectral lines of the Sun.

Meanwhile, Lord Kelvin and Helmholtz both thought the age of the Earth was 20 to 22 million years at most, based on their ideas of the interior heat mechanisms of the Sun. But geologists and biologists of the day were coming up with vastly different figures for the age of the Earth. In the ongoing search for a definitive age for the Earth, Lord Kelvin investigated geomagnetism, hydrodynamics, the shape of the Earth and the geophysical determination of the Earth's age. He soon found himself at the center of a controversy concerning the age of the Earth because his estimated age of the Sun at only 20 million years was not nearly long enough for a gradual process of biologic evolution to take place on Earth, whereas geologists such as James Hutton and Charles Lyell postulated a much longer span of history for the Earth. Darwin, in formulating his evolutionary theory (see Chapter Six), was using Lyell's figures that Earth's geological history spanned at least 300 million years. More recent, 20th-century understanding of the heat mechanisms of the Sun supports Darwin rather than Kelvin.

GAUGING THE EARTH'S AGE

Earth scientists were an entirely different breed from astronomers. Although miners and engineers had long made a study of the ground we walk on, unlike astronomy, geology as a science did not begin to develop until the 18th century and did not reach full bloom until the 19th.

The 1700s had closed with a great controversy raging among geologists, every investigator lining up with one faction or the other, either the neptunists or the plutonists. The captain of the neptunists was Abraham Gottlob Werner (1750–1817), a prominent German geologist who maintained that all Earth's layers were laid down as sediment by the waters of a primeval flood. Chief of the plutonists was a Scottish geologist, James Hutton (1726–97), who contended that the chief driving mechanism in the Earth's formation was heat within, which periodically broke through the crust in the form of volcanoes.

Of the two schools, the plutonists were the more radical. The neptunists saw the Earth's history as a once-and-for-all occurrence, a giant flood (similar to the biblical story of Noah) that laid down the Earth's crust in its current form. This fit best with a literal interpretation of the biblical story of creation—from which scholars concluded that the Earth could be no more than about 6,000 years old. Hutton and the neptunists, on the other hand, maintained that Earth's history had seen a long, slow, continuing process over greatly extended periods of time. They thought that the same forces now evident at work on the Earth's surface had always been at work, forming, wearing down and reforming, over and over—the molten lava from within pushing up, forming new, crystalline rocks such as basalt and granite and tilting the sedimentary layers of rock at the surface. This point of view was considered radical and rationalist (a point of view accepting reason as the only authority) and was looked on with much skepticism at the outset.

The great French comparative anatomist, Georges Cuvier (1769–1832), numbered among those who objected. Cuvier saw evidence of a series of catastrophes in Earth's history, during which all species were wiped out and after which new rock strata were laid down. The most recent of these catastrophes, he said, was the flood described in the biblical tradition.

Swiss-American geologist Louis Agassiz (1807–73) held an independent catastrophe theory that Earth had suffered an ice age—as many as 20 of them, in fact—counting as evidence some of the current movements and behavior of glaciers and glacial evidence in areas where no glaciers now exist. Although the ice age theory was rejected at first, it gradually became accepted as evidence piled up.

Often called the Heroic Age of Geology, the period from 1790 to 1830 was a time when geology experienced considerable influence from the Romantic movement, which embraced Nature (with a capital "N") and encouraged exploration of exotic lands. Romantics gladly turned their backs on polite, stifling society to explore the wilds of uncultured Nature, wild and unspoiled by human touch. Travel to scenically dramatic locales became popular, and for the scientists who answered the call of this yearning, field work in a wide variety of terrain put them face to face with formations and strata they might otherwise never have encountered. Whereas much of geology, up to this time, had been the study of mineralogy and the identification of isolated rock specimens, it now had become a visionary science

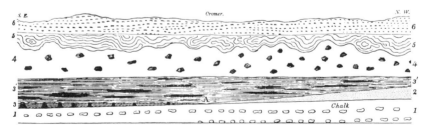

Geologists in the 19th century studied strata like these in the Norfolk cliffs to determine the history of the Earth. (The Geological Evidences of the Antiquity of Man, by Sir Charles Lyell, 1863)

that read strata for their saga of revolutions, of decay and restoration, of great, warring Earth forces.

The old guard, of course, persisted. These were the geologists who continued the practical surveying tradition of the past and who concerned themselves with the professional standing of their science, the collection of evidence and the sound formulation of theory. The Romantic group, often in conflict with the more traditional geologists, saw themselves as knights in the pursuit of Truth, prepared to face its radical consequences and committed to the exploration of Nature.

These two mindsets didn't necessarily divide along ideological lines, however, or substantially affect the methods used. The conservative religious and anti-Revolutionary frame of mind that characterized the years following the French Revolution enforced on geology a strict adherence to empiricism—that is, the support of theory with meticulous evidence. The result was that even those influenced by free-wheeling Romanticism examined strata and collected specimens in much the same way as their col-

Charles Lyell, author of uniformitarianism in geology (The Bettmann Archive)

leagues. But their broader vision and far-ranging field work contributed much to the growth of the discipline, while at the same time they were held in check and in tension by conflicts with their less flamboyant colleagues.

By 1830, a much larger body of facts existed from which to draw theory, and the cause of Hutton's uniformitarianism had attracted a wealthy young Scottish lawyer who was more drawn to geology than to law. Though he studied geology at Oxford under a neptunist, Charles Lyell traveled widely in Europe and had the opportunity to examine many strata of rock for himself, and he soon came to the conclusion that Hutton was right, that the forces that had fashioned the history of the Earth were uniform over time and were the same as those still effecting erosion and sedimentation, heating and cooling in the modern era. He was also widely read—more so than Hutton—and, though he made no discovery, nor fashioned any theory himself, his great gift was that he brought many facts together in one work.

Only the geological forces now in effect should be considered as explanations for past history, he maintained, and very long periods of time should be assumed. He wrote: "Confined notions in regard to the quantity of past time have tended more than any other pre-possessions to retard the progress of geology and until we habituate ourselves to contemplate the possibility of an indefinite lapse of ages . . . we shall be in danger of forming most erroneous views in geology."

In 1830, Lyell published the first volume of *The Principles of Geology: Being an Attempt to Explain the Former Changes of the Earth's Surface by Reference to Causes Now in Operation.* One copy was destined to become well thumbed as it began a journey the following year aboard the HMS *Beagle* in the most famous voyage in the history of science. Its owner was Charles Darwin.

PART TWO

THE LIFE SCIENCES

DARWIN AND
THE *BEAGLE*'S BOUNTY

*E*volution, as an idea, was in the air at the beginning of the 19th century. Geologists were debating the issue of evolution of the Earth. And the discussion naturally went hand-in-hand with the question of biological evolution. Fossils raised questions. Vestigial organs tweaked the imagination. But it was an explosive issue. For this era it was almost as dangerous to put forth as Copernicanism had been in Galileo's day. Certainly, ideas about evolution were bound to cause a furor.

Lyell himself was drawn to the idea, but he preferred, at least at first, to let it alone. In 1836 he wrote to the astronomer John Herschel:

> *In regard to the origination of new species, I am very glad to find that you think it probable that it may be carried on through the intervention of intermediate causes. I left this rather to be inferred, not thinking it worth while to offend a certain class of persons by embodying in words what would only be a speculation.*

The volatility of the subject was not lost on Charles Darwin (1809–82), who, in any case, set out at the beginning of his career with no aim to prove any such point.

VOYAGE OF THE *BEAGLE*

Initially, Charles Darwin hadn't planned to be a biologist. His grandfather, Erasmus Darwin, had been a biologist of sorts and had even offered a theory of evolution, but professionally was a physician. (Charles's other grandfather, Josiah Wedgwood, was a porcelain manufacturer with an interest in chemistry. Both grandfathers had been core members of a scientific philosophical society known as the Lunar Society.) Charles's father was also a physician, and Charles initially planned to follow the family tradition, but quickly discovered he had no stomach for it. He planned instead to train for the clergy, but at Cambridge University his love for rambling alone outdoors found a suitable outlet in botanical field trips. He also became friends with his botany professor, John Stevens Henslow, to whose home he was often invited for dinner and conversation. "His knowledge," Darwin later wrote, "was great in botany, entomology, chemistry, mineralogy and geology. His strongest taste was to draw conclusions from long-continued minute observations." During their long talks Darwin soaked up both content and method. Henslow was impressed with the enthusiasm and ability of his young student, and, hearing of an opportunity for a naturalist to sail under Captain Robert FitzRoy on the HMS *Beagle*, he didn't hesitate to recommend young Darwin.

The *Beagle*'s mission, commissioned by the British Admiralty, was a five-year voyage to map the coasts of Patagonia, Tierra del Fuego, Chile and Peru; to fix longitude; and to establish a chain of chronological calculations around the world. It was customary to take a naturalist along on voyages of this kind, if for no other reason than to provide intelligent and gentlemanly company for the ship's captain.

The *Beagle* set sail on December 27, 1831. The accommodations were cramped—a shared cabin with the captain, a moody man at best—and there was scant room for the equipment Darwin needed for his work. In the tiny, mahogany-lined compartment, he slept in a hammock that swung mercilessly with every pitch of the ship, and he was plagued with seasickness throughout the voyage. "The absolute want of room is an evil," he wrote dismally in his journal at the outset, "that nothing can surmount."

With him Darwin took four books: the Bible, a copy of Milton, Alexander von Humboldt's account of his exploration of Venezuela and the Orinoco basin, and—unquestionably most important to his scientific future—Volume One of Lyell's *Principles of Geology*. On arrival in Montevideo on the eastern shore of the South American continent, Darwin found Volume Two waiting for him, thoughtfully having been sent by Henslow, to whom the *Beagle*'s naturalist kept up a continual flow of reports (many of them read by Henslow at meetings of the Philosophical Society of Cambridge). Volume Three awaited the docking at Valparaiso, on the other side of the continent.

On the high seas, the voyage may have been a nightmare, but the shoreside opportunities for exploration and observation were a naturalist's paradise. On land Darwin was in his element. With clear, evocative prose, he rose to the occasion in his journals (which were published in five volumes in the 10 years following his return). Off the shore of Tenerife, he wrote:

. . . the air is still & deliciously warm—the only sounds are the waves rippling on the stern & the sails idly flapping around the masts . . . The sky is so clear & lofty, & stars innumerable shine so bright that like little moons they cast their glitter on the waves.

On landing, while FitzRoy set up his observatory to take the measurements the Admiralty had sent him to make, Darwin hiked inland or explored the coast, accompanied by interpreters and, sometimes, other ship's officers. He was impressed by the primeval force of nature, lush undergrowth, exotic birds and animals, and, on the coast, brightly colored sponges and intricate tropical corals. South America held a vast store of plants and animals that Darwin had never seen before: wild llamas in Patagonia, giant tortoises in the Galápagos Islands, pansy orchids in Brazil, fossil seashells high in the Andes, corals in the Indian Ocean. He sent hundreds of specimens home to Henslow and made copious notes and drawings.

In the Galápagos, he was struck in particular by a series of finches (now known as "Darwin's finches") living on widely spaced isles that varied in many ways from those found on the mainland. Similar in size and color, 13 different species had developed with different beak shapes, each adapted, apparently, to its unique feeding niche. A seed-eater had a beak that worked well for cracking seed hulls. Another, on a different island where seeds were not available, had a long, sharp beak that worked well for eating its diet of insects. Another, a vegetarian finch, had a short, fat beak that could pluck

Darwin in the field on the Galápagos Islands, measuring the speed of an elephant tortoise   (The Bettmann Archive)

the buds and leaves it fed on. And so on. Darwin was deeply impressed, he wrote later in his *Autobiography,* by the manner in which the species in the Galápagos "differ slightly on each island of the group, none of these islands appearing to be very ancient in a geological sense. It was evident that such factors as these, as well as many others, could be explained on the supposition that species gradually became modified; and the subject haunted me."

Lyell's second volume of *The Principles of Geology* had already begun to raise some of these issues. He had made a study of the geographic distribution of plants and animals, and he had formed the theory that each species had come into being in one center. "Similar habitats on separate continents," he wrote, "seemed to produce quite different species" which were well equipped to survive in their particular habitats. Here he applied his uniformitarian concepts to biology. New species had constantly emerged, he said, through the history of the Earth; others had become extinct along the way. Since geologic processes were constantly in a state of flux, as they were still, the inception and extinction of species constantly took place in response. A highly successful species might crowd out others in competition for food in the same habitat, resulting in extinction of some species. But Lyell stopped short of "transmutation" of species. A new species might come into existence, but it would not change, or evolve, over time.

PRE-DARWINIAN EVOLUTION

Biology had reached this point by stages. In 1686 John Ray had defined the modern concept of species, based on descent from a common ancestral type, and the Comte de Buffon clarified this in 1749: A species is a group of interbreeding individuals who cannot breed successfully outside the group. But these early biologists assumed that species had not changed since the beginning of time; they were locked into a concept known as the Great Chain of Being, with all species part of an unbroken chain from the lowliest creature at the bottom to humans and angels at the pinnacle. All species were fixed, having the same characteristics they had always had since the beginning of time. And in the 17th and 18th centuries most scientists had subscribed to an idea called preformation that assumed that every adult was already preformed either in the egg or in the sperm (depending which of two warring factions on this point they subscribed to). This theory was a roadblock to the development of the theory of evolution, but it was soon replaced by the theory of epigenesis, established by Caspar Wolff, who saw that an embryo develops not from a tiny preformed creature, but from undifferentiated tissues. The emergence of the theory of epigenesis was the first major step necessary to open the door to an evolutionary theory of the origin of species.

The second factor to prepare the way toward the end of the 18th century and the beginning of the 19th was the understanding of fossils and the fact that they represented (see box) the bones of species that no longer existed.

And the third precondition to emerge was the growing recognition, also toward the end of the 18th century and the beginning of the 19th, that the Earth was very, very old. In fact, James Hutton and Charles Lyell maintained that the formations present in the Earth's crust could only have been produced over great stretches of time. These stretches of time were sufficiently vast (Lyell estimated 240 million years) that evolutionary changes in species might not be observable in living species. (This is in fact the case: We generally cannot observe evolutionary changes at the species level within a lifetime.)

So, by the beginning of the 19th century many scientists had begun to entertain ideas of some form of evolution. Jean-Baptiste de Monet, Chevalier de Lamarck, who was one of the first evolutionists, correctly proposed that species evolve in response to the environment. He also thought, however, that acquired characteristics could be inherited by offspring, an idea with which most scientists of his day and since have disagreed. For instance, a giraffe, he said, could make its neck longer by stretching toward the leaves high in the trees and could pass this acquired trait on to its offspring.

THE ORIGIN OF SPECIES

On his return from the *Beagle*'s voyage in 1836, Darwin was troubled by his observations, and yet he was not prepared to draw conclusions about evolution of species. Finally in May 1837 he began a notebook on evidence for transmutation. Then in 1838 he came across *An Essay on the Principle of Population*, a book published in 1798 by English economist Thomas Robert Malthus (1766–1834). Not about biology at all, Malthus's essay proposed that human populations increase geometrically (e.g., 2, 4, 8, 16, . . .), while means to support them increase only arithmetically (e.g., 1, 2, 3, 4, 5, . . .). So, he said, natural selective forces—such as overcrowding, disease, war, poverty and vice—take over to weed out those that are less fit. Those that are fittest survive.

In his notebooks, Darwin had used the words *descent* and *modification*. Now, thanks to Malthus, he had a new term to describe the process for which he had seen evidence: *selection*. He now wrote: "One may say there is a force like a hundred wedges trying [to] force every kind of adapted structure into the gaps in the economy of nature, or rather forming gaps by thrusting out weaker ones."

COPE AND MARSH:
RIVAL BONE HUNTERS

A group of bone searchers sponsored by Othniel Marsh. In the 1880s Marsh and Edward Cope raced each other to see which of them could dig up the most fossil bones in the mid-western and western United States. (Yale University Archives)

Dinosaur fossils—huge fragments of fossilized bone from the skulls and skeletons of creatures that died millions of years ago—offered one of the most dramatic types of evidence for Darwin's contentions that the Earth was very old and that species had changed over time. And during the mid-1800s the American West and Midwest proved to be a bone hunter's paradise for paleontologists seeking dinosaur bones. Two of the most successful and most respected of the American bone hunters were Edward Drinker Cope (1840–97) and Othniel C. Marsh (1831–99). Scouring through the West with ferocious intensity, the two men collected an astounding array of dinosaur finds, published hundreds of papers and enriched the collections of many museums. None of this was done in a cooperative spirit, however. The two men hated each other and it was often said that they mounted their western expeditions more like raiding parties than carefully planned scientific expeditions.

Neither could be considered "innocent" in the conflict. Cope was wealthy, egotistical, belligerent, and often—as Marsh shouted accusingly—dishonest. Marsh was wealthy, egotistical, belligerent and often—

Ultimately, Darwin developed the idea of natural selection in species, a concept that is often referred to as "survival of the fittest." That is, those individuals of a species that are fittest to reproduce are the ones who successfully pass on their traits to later generations (even though these as Cope shouted accusingly—dishonest! Many of their contemporaries were happy that the two men hated each other so much, and spent so much time trying to best the other: The fact that they focused their unpleasant personalities and unscrupulous activities on each other kept them pretty much out of everybody else's hair.

At least, some of the time. Once, when he was visiting Cape Cod, Cope watched from a distance while a team of scientists from Harvard University worked all day to cut away the decaying flesh of a gigantic whale that had washed ashore. The huge skeleton was being prepared to be sent to the Agassiz Museum at Harvard. After the team had finished their work, loaded the bones on a railroad flatcar and wearily left for home and rest, Cope came out of hiding and bribed the railway stationmaster to switch the destination on the shipment card from the Harvard museum to his own museum in Philadelphia, Pennsylvania. It was a few years before the Harvard team discovered how their gigantic skeleton had become lost.

Most of Cope's raiding, though, took place at Marsh's expense, and Marsh paid him back in kind. Typical of their behavior toward one another was the time when Cope made a deal with an amateur bone hunter for some important bones Cope wanted. Cope promised the man he would send him a check and collect the bones—and then proceeded to write a paper on them. Marsh, though, also heard about the amateur's find and sent a message to the man, pretending to be a Cope associate and canceling the deal. Then Marsh bought the specimens himself.

The rivalry became so absurd at times that it threw American paleontology into confusion. Not the most careful of scholars, both rushed into print often, and since they often covered the same territory, they often found the same species at the same time, which they both then went about naming—resulting in many dinosaurs with two different names, and leaving the mess of sorting the whole thing out to later paleontologists.

Although they conducted much of their science in a decidedly ungentlemanly and often unscientific manner (Cope often dynamited his research sites after he was finished with them, so no one else could come along and get to something he might have missed!), the two buccaneering paleontologists together discovered and named over 1,718 new genera and species of fossil animals from the American West.

"fittest" may not necessarily be the strongest or the best). For Darwin, the idea was beginning to gel. In the true tradition of Francis Bacon, he had set out with no preconceived ideas about what he would find, had collected voluminous evidence and now had spent time reflecting on it and analyzing

Charles Darwin at the age of 40, at the time he was working on barnacles (Courtesy National Library of Medicine)

it. In 1842 he began to write a 1,500-word synopsis. By July 1844 he had written a 15,000-word essay, which he showed to his friend, botanist Joseph Dalton Hooker (1817–1911).

That same year, Robert Chambers published a book called *Vestiges of the Natural History of Creation* that set forth a theory of evolution, which intrigued Darwin. Although Chambers's book had popular appeal, however, Chambers offered no explanation for evolution and was not careful, making many errors that discredited him with most scientists.

Determined to make certain that his own work was so thorough and well documented it would be irreproachable, Darwin embarked on an eight-year study of fossil and living forms of barnacles. He examined 10,000 specimens, publishing four monographs on his work between the years 1851 and 1854. It was exhausting, meticulous work, and tedious, but, at the end of it, Darwin felt he had the right to call himself not just the collector and observer he had been aboard the *Beagle*, but now truly a trained naturalist.

During this period he came upon a key idea, the idea of divergence. How did differences between varieties become so pronounced that they became differences that separate species—differences that made inter-breeding impossible? "The more diversified the descendants from any one species became in structure, constitution and habits," he wrote, "by so much will they be better enabled to seize on many and widely diversified places" in nature's structure. And that very diversification, that branching off, took them further and further from the original stock.

By now he had become good friends with Lyell, who at first rejected Darwin's ideas about evolution. Now Lyell, Hooker, and Darwin's brother, Erasmus, all encouraged him to begin writing a book that would set out his theory definitively and provide all the documentation he could amass to support it. And Darwin began writing. "I am working very hard at my book," he wrote to a cousin in February 1858, "perhaps too hard. It will be very big; and I have become most deeply interested in the way facts fall into groups. . . . I mean to make my book as perfect as I can. I shall not go to press at soonest for a couple of years."

LETTER FROM MALAYSIA

But Darwin's "big book" would never be completed. With 250,000 words written, on June 18, 1858 he received an astounding letter from a fellow

Alfred Russel Wallace, who, based on his own observations, independently came up with many of the same ideas about evolution as Darwin (Courtesy National Library of Medicine)

naturalist working in Malaysia. Would Darwin read the enclosed essay and, if he thought it worthy, pass it on? The essay, which treated "the tendency of varieties to depart indefinitely from the original type," by Alfred Russel Wallace (1823–1913), delineated the same theory of evolution Charles Darwin had been working on for 20 years! The resemblance was uncanny.

Completely dismayed, Darwin sought the advice of Lyell and Hooker, who suggested he explain the predicament to Wallace and propose a common announcement of the theory, taking joint priority. Darwin followed the suggestion and was delighted to hear from Wallace that he was most happy to make a joint announcement. Actually, the younger man had spent much less time and much less care developing the theory, and he

Darwin in his sixties, about 15 years after publication of On the Origin of Species
(Courtesy National Library of Medicine)

profited from Darwin's stature. So in 1858 Wallace's essay appeared in the *Journal of the Linnean Society* along with a summary by Darwin of his work.

Darwin published his shortened version in book form on November 24, 1859: *On the Origin of Species by Means of Natural Selection, or the Preservation of Favoured Races in the Struggle for Life.* (It's usually known just as *On the Origin of Species* or the *Origin of Species*.) Every copy of the original 1,250-copy printing was sold on the first day. Even in this shortened form it is a very long and persuasive argument for evolution, supported by many, many examples, and it succeeded in convincing a number of biologists of its truth.

DESCENT OF MAN

But, while most scientists accepted Darwin and Wallace's evolutionary theory and the idea of natural selection, the public was slower to go along with these ideas. Many considered Darwinism to be atheistic—and, in fact, Darwin had struggled long and hard with this issue himself. If organisms adapted through the mechanistic process of natural selection, where was the role of God in the creation or continuing development of the world and its creatures?

And some scientists also balked mightily. Richard Owens, a renowned geologist at Oxford, and Louis Agassiz at Harvard in the United States numbered among the most vocal scientific opponents. Adam Sedgwick, an old-school geologist from Cambridge, was another, remarking in 1865 of his colleague Charles Lyell:

> *Lyell has swallowed the whole theory [of evolution], at which I am not surprised, for without it the elements of geology as he expounded them were illogical . . . They may varnish it as they will, but the transmutation theory ends, with nine out of ten, in rank materialism.*

The most troubling point for most people, though, if one species evolved from another, was the idea, assiduously avoided by Darwin in his book, that human beings must have descended from a nonhuman ancestor.

The controversy became hotter and hotter. Apelike cartoons of Darwin appeared in newspapers. Essays and sermons proliferated everywhere. Darwin had never recovered from the untimely death of his daughter Annie and was in ill health, which only became worse as the controversy raged. He was, in any case, of no temperament to combat the furor. T. H. Huxley (1825–95), who had read *On the Origin of Species* with relish ("Why didn't I think of that?" he wanted to know), leaped to the lecture platform in his place. Calling himself "Darwin's bulldog" (Huxley always loved a good intellectual

Cartoons like this one appeared in countless periodicals in Darwin's time. Here, the go-
rilla, pointing to the bearded Darwin, exclaims: "That man wants to claim my Pedigree.
He says he is one of my descendants." To which Mr. Bergh (center), founder of the Society
for the Prevention of Cruelty to Animals, rejoins, "Now, Mr. Darwin, how could you
insult him?" (The Bettmann Archive)

fight), he debated the issue of evolution with zeal wherever and whenever
he could.

Before a packed audience of 700, Huxley met the Bishop of Oxford,
Samuel Wilberforce (known as "Soapy Sam," for his slippery manner and
unctuous speech) in a grand debate before the British Association for the
Advancement of Science. But Wilberforce, probably primed by Owens with
"facts," lost the day before the Victorian audience when he snidely inquired
which of Huxley's parents descended from the apes, his father or his mother.

Huxley knew when he had an edge and he used it. The audience was there
for a show, and the quick-witted Huxley provided it. Better to be descended
from an ape, he replied evenly, than from an educated man who would

*T. H. Huxley, defender
of the theory of
evolution and
Darwin's ideas about
natural selection*
(Courtesy National
Library of Medicine)

introduce such a question into a serious scientific debate. For that, Wilberforce had no retort.

In 1863, Lyell finally took up the cause with a book entitled *Geological Evidence of the Antiquity of Man.* He came out in favor of Darwinism, though a bit more timidly than his friend would have liked. He did indicate that humans or humanlike creatures must have existed for many millennia on the face of the Earth. But he did not say openly that he believed in transmutation or that humans had evolved from apelike creatures.

In 1871, Darwin came out with *The Descent of Man, and Selection in Relation to Sex*, in which he supported the idea that humans descended from pre-human creatures with considerable evidence from his research. Humans possess vestigial points on their ears, he pointed out, and muscles that once obviously were used to move the ears. The "tail-bone" or coccyx at the base of the spine he cited as another example.

Resistance to these ideas continued in some sectors—George Bernard Shaw also opposed Darwinism, subscribing instead to Lamarckianism and a kind of mystical life force. As late as 1936, H. G. Wells wrote *The Croquet Player* examining the unease of the civilized English gentry when faced with the knowledge of the "brute within." But most people were won over by the end of the 19th century, except for those who opposed evolution as antithetical to the biblical story of creation.

MORE PROOF

There were still some problems, though. Wouldn't a variation in one individual, however successful, become blended into intermediate, less successful versions as it was passed on to later generations of offspring? How could natural selection ever have the opportunity to operate on successful variations?

A lone Augustinian monk from Austria named Gregor Mendel, running experiments with dwarf and tall strains of peas in the monastery garden, came up with the proof in the 1850s and 1860s. Mendel discovered that for

THE NEANDERTHAL MYSTERY

When workers unearthed a partial humanlike skull and skeleton from a limestone cave near the village of Neander, Germany in 1856, they also unearthed a mystery that continues today.

The renowned French paleontologist Georges Cuvier (1769–1832) had argued hotly that no fossils of ancient humans would ever be discovered for the simple reason that they didn't exist. Cuvier was what we would call today an anti-evolutionist, and most of the scientific world before Darwin agreed with him.

To whom did those strange bones belong, then—and strange they were, with a heavy skullcap and nearly a dozen and a half curiously "deformed" skeletal parts. The most obvious guess that came to the first German professors who examined the bones was that they were definitely not German, but probably belonged to some much wilder and older northern tribe. Rudolf Carl Virchow [FEER koh], Germany's most respected anatomist, didn't believe they were even ancient but thought they most likely belonged to an unfortunate cripple with a badly deformed skull who also suffered from arthritis. Another German anatomist, perhaps wondering about the consequences of such a strange-looking individual wandering around the German countryside, suggested that such a person might even have been mentally "abnormal" and lived as a hermit in the cave in which the bones had been found.

Today, after Darwin, and now that more and more of the so-called Neanderthal skeletons have been found in places ranging as far apart as North Africa, France, Israel and China, we know that the Neanderthals, like many other fossil human finds, do have a place in the story of human evolution. But exactly what place still is not certain.

each trait an offspring apparently inherited factors from *both* parents equally. Most important, he found that those factors are not blended but remain distinct and can, in turn, be passed on, unblended, to subsequent offspring. This meant that natural selection would have much more time to operate on any variation in a trait. Unfortunately, Mendel's work did not become widely known until it was rediscovered at the beginning of the 20th century.

Evolutionary theory has not stopped growing; many 20th-century biologists have added to Darwin and Wallace's original theory and have incorporated newer knowledge about the genetic mechanics of mutation and heredity. Emerging theories about the history of the Earth have also

---

Early portraits of Neanderthals, based primarily on that first find in the Neander Valley, portrayed a stooped, hulking, shambling creature with brutish intelligence—the "ape man" of countless bad movies. For many in the 19th century and well into the 20th, this portrayal of Neanderthals as the direct ancestor of *Homo sapiens* raised deeply troubling questions about "the beast within" us all. Ironically, some of the earliest guesses about the first Neander Valley specimen's serious arthritic deformity, if not about its age, have turned out to be correct. More recent finds have indicated a much less stooped stature and appearance.

Neanderthals roamed the world beginning about 130,000 years ago and disappeared about 35,000 years ago. They were short and squat; the men probably averaged only a little over five feet, and the women were slightly shorter. They walked erect, made tools and lived in social groups. Ceremonial objects occasionally found with the dead also suggest that they may have developed a fairly advanced religious system.

But the long-standing mystery remains: What relationship did Neanderthals have with modern humans, and why did they disappear from the face of the Earth? Through the end of the Victorian era most scientists believed that Neanderthals were our direct ancestors, but more recent findings in the 20th century show that our direct ancestors—the Cro-Magnons—lived at the same time as the Neanderthals. As a result some serious questions have cropped up about the Neanderthal's direct relationship to us.

Perhaps the biggest mystery, though, about these intriguing people is what happened to them to have caused their complete disappearance about 35,000 years ago. Did our direct ancestors, Cro-Magnon, systematically and violently kill off their Neanderthal neighbors, as some have suggested? Did Cro Magnon and Neanderthal mix and inter-breed, gradually merging into one population, as others argue? Or did they, like many other populations before them, simply lose out in the long-term battle for survival, gradually diminishing until extinction finally overcame them?

contributed to modern evolutionary theory, including the idea that periods of stability on the Earth's surface have been "punctuated" by times of extreme change, most recently proposed by Stephen Jay Gould and Niles Eldredge.

Charles Darwin died of a heart attack at his home on April 19, 1882. His pallbearers included those who had stood by his side in science—Huxley, Wallace, Hooker and Sir John Lubbock—and he was buried at Westminster Abbey with Sir Isaac Newton and Sir Charles Lyell. Darwin, however, was never knighted by the British monarch and no statue of him joins the other giants of his century in Madame Tussaud's wax museum in London—perhaps as a caution against vandalism. For this quiet, orderly man with a great respect for natural law stirred up mammoth controversies in his time that remain for some people to this day. He has come to symbolize the intellectual doubts of his century—although, of course, he in no way created them single-handedly. Charles Darwin's ideas about evolution weren't new and they weren't complete. But they did spawn a revolution, altering for all time humanity's understanding of itself and its own place in the natural universe.

FROM MACRO TO MICRO: ORGANS, GERMS AND CELLS

*E*volution looms like a giant in the history of 19th-century life science, but it was not the only area of research that saw important new developments. François Magendie and Claude Bernard in France pioneered in the fields of experimental physiology, pharmacology and nutrition. Ivan Pavlov in Russia gained important insights about the functioning of the brain. Louis Pasteur, arguably one of the three most important biologists in the history of science, established insights into the mechanisms that make organisms sick. And Theodor Schwann and Matthias Schleiden together discovered that for living organisms nature had another basic building block above the level of molecules: the cell.

EXPERIMENTAL PHYSIOLOGY

Science benefits from no single trait more than a strong spirit of inquiry and zest for knowledge. But, while François Magendie (1783–1855) opened up many avenues of research and pioneered in important areas of experimental physiology, his lack of sensitivity about his experimentation raised enormous controversy in his time and would probably raise even more today. Recognizing that often the most complete knowledge came from seeing living tissue as it functioned—especially in the case of experiments on motor nerves—he did most of his experiments on living organisms (usually dogs), which brought the antivivisectionist activists up in arms against him. Described as irascible and ruthless, he depicted his own activity as an "orgy of experimentation." He often experimented when it wasn't necessary, pushing beyond the limits of decency, and he earned a rather unsavory reputation.

However, Magendie did some good science that solidly expanded knowledge, even if his methods were unsavory. He demonstrated the functions of the spinal nerves, showing that the anterior nerve roots of the spinal cord are motor (convey messages to enact movement) and the posterior nerve roots are sensory (conveying messages to the brain regarding sensations). He also investigated the mechanisms of blood flow, as well as swallowing and vomiting. And Magendie introduced the use of strychnine, morphine, brucine, codeine and quinine, as well as the compounds of iodine and bromine, into medical practice. What he may have done best, however, was pass on his gusto for research to his students without infecting them with his disregard for his research subjects.

Claude Bernard (1813–78)—also known for his work in experimental physiology—became Magendie's assistant in 1848 and succeeded him as professor in 1855 when Magendie died. Bernard is considered to be the founder of experimental physiology. Unlike Magendie, Bernard planned his experiments carefully and integrated them, and his contributions to physiology are extensive.

Bernard studied digestion through fistulas, openings introduced through the wall of the stomach, and he was able to gain considerable insight into the chemical phenomenon of digestion. He was interested in the balance of

*Claude Bernard,
founder of
experimental
physiology* (Courtesy
National Library of
Medicine)

chemicals in the body—always objectively seeking biochemical explanations when others would invoke vitalism or mysticism—and he discovered the role of the pancreas in the digestion of fat and the glycogenic function of the liver.

Bernard was able to prove that, contrary to prevailing opinion, animal blood contains sugar even when sugar is not present in the diet. He found that animals convert other substances into glucose, or sugar, in the liver; using experimental methods, he was able to track down the presence of glycogen, the chief animal storage carbohydrate, in the liver. He also discovered the synthesis of glycogen, known as glycogenesis, and the breakdown of glycogen into glucose, known as glycogenolysis. (Glycogen is stored in the liver until it is needed for release into the bloodstream to maintain the glucose level.)

Bernard also detected how certain nerves (called vasomotor nerves) control the flow of the blood to the skin, causing capillaries to dilate when the skin needs cooling and constrict when the body needs to conserve warmth.

In a study of carbon monoxide poisoning, Bernard was the first to provide a physiological explanation of the process by which a drug affects the body. Carbon monoxide, he realized, replaces oxygen molecules in the blood and the body cannot react quickly enough to prevent death by oxygen starvation.

Unfortunately for Bernard, despite his greater care, his wife abhorred his vivisections and argued with him about his work constantly. She made large contributions to antivivisectionist societies and legally separated from Bernard in 1870. Beset with financial problems, ill health and family quarrels (his two daughters also campaigned against his work), Bernard once described the life of science as "a superb and dazzlingly lighted hall which may be reached only by passing through a long and ghastly kitchen."

The French Académie des Sciences awarded the grand prize in physiology to Bernard on three different occasions, and he became a senator in 1869. He was the first scientist to whom France gave a public funeral.

PAVLOV AND THE BRAIN

For early philosophers, the brain represented the seat of the rational soul, which people possessed and which plants and animals did not. In the Middle Ages, scholars thought that imagination, reason, common sense and memory each resided in one of four chambers, or ventricles, in the brain.

The Renaissance brought new methods of examination, and dissection unmasked many of the mysteries of animal (and human) physiology. Scientists even studied the convolutions on the surface of the brain. But as late as the 17th century the French philosopher René Descartes [day KART] still clung to many medieval ideas, even though he tried to apply Newtonian principles of mechanization to physiology. Descartes identified the pineal body at the base of the brain as the seat of the human rational soul, which he said received sensory messages and responded to them by controlling the flow of animal spirits through hollow nerves. In this way, the rational soul directed the movement of muscles. This was one of the first primitive explanations of reflex action.

The convoluted surface of the human brain led some anatomists, among them the 17th-century English anatomist Thomas Willis (1621–75), to think that particular functions were seated in various areas of the convolutions. The practice of phrenology developed from this idea and bloomed at the end of the 18th century as "the science of the mind." Phrenologists claimed that they could construct a map of the surface of the brain, locating particular areas that controlled such qualities as acquisitiveness and destructiveness (over the ear), moral qualities such as benevolence and spirituality (near the top of the head), and social or domestic inclinations, such as friendship (near the rear of the brain). Bumps on the surface of the skull were said to correlate with relative development of these various areas of the brain, and phrenologists asserted that they could "read" these bumps and analyze personality and character in this way. Phrenology generally was discredited as a science by the 1840s, although it lingered as a popular pseudoscience well into the 1920s.

However, experiments performed by German physiologists Julius Hitzig [HIT sikh] (1838–1907) and Gustav Fritsch (1838–1927) were the first to show that different parts of the brain do, indeed, control different functions. Hitzig performed experiments on living dogs, showing that stimulation of a particular region of their brains caused contraction of certain muscles. He also showed that damage of particular portions of the brain caused certain muscles to become weakened or paralyzed.

The Scottish neurologist David Ferrier (1843–1928) also did experiments along these lines, on primates as well as dogs, and succeeded in showing that some regions of the brain (motor regions) controlled movement of the muscles and other organs, while others (sensory regions) received sensations from the muscles and other organs. Like Bernard, Ferrier ran up against opposition from animal rights activists of the day, who accused him of cruelty to animals. But he succeeded in showing in court (1882) that valid justification existed for the experiments, which established crucial knowledge about how the brain functions.

Ivan Pavlov (1849–1936), the Russian physiologist, is best known for his experiments with automatic reflexes and animal behavior.

Pavlov's early work, for which he won the 1904 Nobel Prize for physiology or medicine, investigated the physiology of digestion and the autonomic nervous system. In a dramatic experiment he surgically bypassed the stomach of a dog so that food eaten by the dog would go down its esophagus but never reach its stomach. Strangely enough, the stomach's gastric juices flowed anyway, just as if it had received the food. Pavlov concluded that nerves in the mouth must convey a message to the brain which then triggers the digestive reaction. He was able to show this further by severing certain nerves and observing that now the dog might eat and (without the bypass) food would reach the stomach, but no gastric juices flowed to digest the food. The brain had not received the message.

Pavlov is best known, however, for his later work on conditioned reflexes (which had a marked influence on physiologically oriented psychology in both Germany and the United States). Pavlov found that, since a dog salivates when it sees food, he could substitute another stimulus—a bell, for example—for the food and, as long as the primary stimulus (food) was associated with the secondary stimulus (bell) during a training period, the dog would salivate when it heard the bell. Pavlov supposed that new circuits of nerves developed in the cortex of the brain to allow these involuntary "conditioned" reflexes to occur.

THE BIRTH OF CELL THEORY

Today the idea that living things are composed of cells seems the most basic of concepts, but before the 19th century, most biologists analyzed the makeup of animals and plants no further than the recognition that living organisms were composed of tissues and organs. As early as 1665, Robert Hooke had observed cells in cork through a microscope—and, thinking they looked like the tiny rooms called cells in a monastery, he gave the tiny structures their name. But he had no idea that what he was looking at was part of a fundamental principle of life.

Most early observations of cells were done on plants, which are easier to see because they have cell walls, which are thicker than the cell membranes found between animal cells. And, with the improvement of microscopes and staining techniques (to bring out the various structures and make them more visible), scientists made more and more observations. But even in 1831, when Robert Brown discovered a small, dark structure at the center of cells and named it the "nucleus" (from the Latin word for "little nut"), neither he nor anyone else yet understood the significance of these tiny structures. In 1835 a Czech scientist named Jan Evangelista Purkinje [POOR kin yay] pointed out that certain animal tissues such as the skin are also made of cells. But no one paid much attention, and he did not push the point to state a full-blown theory.

Just three years later, though, in the first part of a stunning one-two punch, Matthias Jakob Schleiden [SHLY den] (1804–81) set forth the surprising idea that all plant tissues were actually made of cells—that these were nature's building blocks for all plant life. The follow-up came the next year, when Theodor Schwann [SHVAHN] (1810–82) suggested that all animal tissues were also composed of cells and that an egg is a single cell from which an organism develops, that all life starts as a single cell. Since each contributed an essential part of the picture, both Schleiden and Schwann usually get the credit for the cell theory.

Others refined the details of the theory. Schleiden thought that new cells formed as buds on the surfaces of existing cells (which Mendel's nemesis, Karl Wilhelm von Nägeli, showed was not true). But cell division remained a mystery for many years. In 1845 Karl Theodor Ernst von Siebold extended the cell theory to single-celled creatures (although he thought that multi-cellular organisms were made from single-celled creatures). And in the 1840s Rudolf Albert von Kölliker demonstrated that sperm are cells and that nerve fibers are parts of cells. The cell theory rapidly became one of the main foundations of modern biology.

Theodor Schwann, who receives credit (along with Matthias Jacob Schleiden) for the development of cell theory (Courtesy National Library of Medicine)

Rudolf Virchow, founder of cell pathology   (Parke-Davis, Division of Warner-Lambert Company)

VIRCHOW AND CELL PATHOLOGY

Rudolf Carl Virchow was a man of passions. As a young physician in what is now Poland, he spoke out against the social conditions where he was looking into the causes of a typhus epidemic. It was the time of revolutionary uprisings—the year was 1848, the great year of revolutions—and the ultra-conservative government of Prussia sternly put down all dissent. Virchow lost his university professorship for his stand.

But three years earlier, in 1845, he had been the first to describe leukemia, and now, in semi-retirement, he had time to reflect on his ideas about the causes of disease. The following year he found a position in Bavaria at the University of Würzburg.

Seven years later, he returned to Berlin as professor of pathological anatomy, a field that he pioneered, and in 1858 he published *Cellular Pathology*, a book in which he established that cell theory extended to diseased tissue. The first cellular pathologist, he showed that diseased cells descend from normal cells. "All cells arise from cells," he was fond of saying, in an implicit repudiation of the idea of spontaneous generation, which maintains that living matter springs out of nonliving matter.

Given Virchow's background in the study of leukemia and cell pathology, it's not surprising that he thought of disease as caused when cells revolt against the organism of which they are a part. As a result, he resisted Pasteur's germ theory. (Though it's possible to see cancer as cells warring on cells, Virchow saw all disease this way.)

PASTEUR'S GERM THEORY

Of Louis Pasteur (1822–95), the prolific science writer Isaac Asimov once wrote, "In biology it is doubtful that anyone but Aristotle and Darwin can be mentioned in the same breath with him." As the founder of germ theory, the instigator of the pasteurization process for sterilizing dairy foods and the inventor of a vaccine for rabies, Pasteur's name has become a household word. Pasteur's enormous achievements resulted from a series of intense controversies in which his ego prevailed (he liked to be right) and his perseverance paid off with major breakthroughs first in understanding the nature of fermentation (which he said was organic; his foes claimed it was inorganic) and the question of the possibility of spontaneous generation (not possible he said; possible, his foes insisted). In each case Pasteur won the day, and in the end emerged a recognition that bacteria can cause disease.

Louis Pasteur, one of the greatest biologists of all time, established germ theory, a breakthrough that held major importance, not only for the study of biology, but also for health science and medicine. (Parke-Davis, Division of Warner-Lambert Company)

Because of Pasteur's success with the tartrate crystals in his first serious project (see Chapter Two), he was already famous in his early thirties, although he was turned down for membership in the Académie des Sciences in 1857. He did, however, accept an appointment as dean of the Faculty of Sciences at the University of Lille. There, the wine and beer industry had a problem with their products going sour, and Pasteur was asked for his help. And so began Pasteur's inquiry into the nature of fermentation, which he found to be a product of a certain type of living organism (a battle he won against Justus von Liebig, who had always maintained that fermentation was a purely chemical reaction, involving no living organisms). The problem was to allow the yeast organisms that produced fermentation in the wine or beer to do their job, but not let the ones that produced lactic acid (which made the beverages go sour) do their job. He suggested heating the wine and beer slightly to kill the "bad" yeast after fermentation was complete and then capping. This gentle heating process to kill undesirable microscopic organisms is now called pasteurization.

Pasteur then turned to other microscopic organisms to examine the question of where they come from. The question of spontaneous generation still haunted the study of biology. Those who believed that organisms had some vitalist essence criticized the experiments intended to show that spontaneous generation could not occur. The environment could not be hostile to life, or of course life would not occur—spontaneously or other-wise. They objected to experiments done the previous century by Lazzaro Spallanzani, in which he heated the air above the flasks of broth in which he intended to show that life could not arise from nonlife. By heating the air, they said, he had destroyed a necessary vitalist principle.

So Pasteur devised a special long-necked flask, which he sterilized. The long, thin, curved neck could be left open to allow oxygen to enter the flask, but the opening was so small that spores floating in would get caught in the bend. It worked. No organisms grew in the flask itself. But he was able to show that spores of living organisms had in fact gotten caught in the crook of the neck. He had put the question of spontaneous generation to rest at last. Now he was admitted to the Académie des Sciences.

When the silk industry in the south of France got into trouble in 1865, its leaders called in the wizard, Pasteur, who quickly identified a microscopic parasite infesting the silkworms and the leaves they fed on. Destroy the infested worms and food, he said, and start afresh. The deed was done, and the silk industry was saved.

Now Pasteur turned his attention to the single greatest achievement of his career: the germ theory of disease. He began to realize that disease was communicable and that the spreading was caused by tiny, parasitic micro-

SHUTTING OUT GERMS

By mid-century, even before Pasteur's discoveries, a few physicians had begun to recognize that doctors could carry disease and infection from one patient to another, although they didn't know exactly what the carrier was.

Ignaz Philipp Semmelweiss (1818–65), a Hungarian physician trained at the University of Vienna and working in hospitals there, became concerned about an unsettling fact: Women giving birth in hospitals were dying in droves of a disease known as childbed fever, while among those who gave birth at home far fewer contracted the disease. Doctors, he became sure, were carrying the disease from patient to patient, and he ordered all the doctors under him to use a strong chemical solution to wash their hands between patients. The doctors grumbled—no doubt unhappy with the idea that they were causing, not curing, disease in their own patients. But they did it, and the incidence of childbed fever dropped off dramatically.

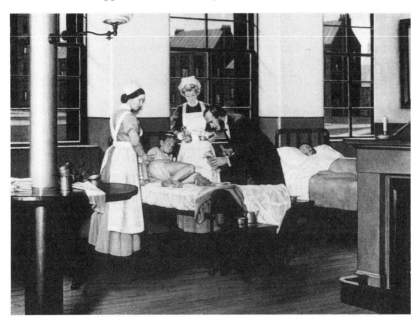

Baron Joseph Lister   (Parke-Davis, Division of Warner-Lambert Company)

organisms he called "germs." (We now know "germs" are bacteria and viruses—see glossary of explanations.)

By understanding this, he realized, the method of disease communication could be stopped. He soon was advising military hospitals about sterile

But in 1849, Hungary staged an unsuccessful revolt against the Austrian Empire (of which it formed a part), and, as a Hungarian, Semmelweiss was forced to leave his post in Vienna. The Viennese physicians went back to their old ways, skipping the unpleasant handwashing routine, and the childbed fever deaths among their patients climbed high again.

Meanwhile, wherever he worked, Semmelweiss continued to insist on his procedure, and he was able to reduce the number of childbed fever deaths among patients under his care to 1 percent. He could show that handwashing worked, but no one had yet shown that it was working because dangerous germs were being destroyed. Ironically, Semmelweiss died of childbed fever from a wound he inflicted on himself while working with a patient, but he had paved the way for the work of another perceptive physician, Joseph Lister.

Joseph Lister (1827–1912) was interested in amputations—a method that promised to save lives in many instances, such as when gangrene had set in. But much too often—in 45 percent of the cases—a physician would complete an operation successfully only to see the patient die from infections afterward. In 1865, he heard about Pasteur's research in diseases caused by microorganisms, and he came up with the idea of killing germs with chemical treatment. He began using an antiseptic solution known as carbolic acid (phenol) to clean instruments in 1867. He also sprayed the air in the operating rooms, and insisted on hand washing and clean aprons. The surgical death rate dropped from 45 percent to 15 percent. Joseph Lister had established antiseptic surgical technique and had killed the germs.

Then an American surgeon named William Stewart Halsted (1852–1922) took the antiseptic concept one step further. Why not just put a shield between the germs carried by physicians or nurses and their patients? So in 1890 Halsted became the first surgeon of major stature to use rubber gloves in surgery. Gloves could be sterilized much more drastically than the skin of the human hand; they could be subjected to high temperatures and caustic chemicals that could eliminate the presence of even the hardiest germs. The wearing of rubber gloves instituted the first use of aseptic surgical technique (in which germs are not just killed in the operation room, as in antiseptic technique, but absent).

Semmelweiss's experience and Pasteur's germ theory had laid a foundation on which Lister and Halsted could build. As a result, physicians changed their operating-room practices and many thousands of lives were saved.

technique—boiling instruments and sterilizing bandages—to prevent infection and, often, avoidable death.

In the 1870s, Pasteur took on anthrax, a particularly deadly, highly communicable disease of domestic animals. In 1876 Robert Koch, a young

ROBERT KOCH:
FINDING CAUSES OF DISEASE

In the early 1870s Robert Koch (1843–1910), a young physician in a tiny town in Germany, responded to an appeal from farmers for help combating the dread anthrax epidemic that was destroying their herds of cattle. Inspired by Pasteur's work with germ theory, Koch [KOKH] was delighted by the opportunity to try to find the cause of a disease instead of just treating its symptoms. He set up a small laboratory in his home, equipped with a microscope, and began investigating blood specimens from infected cattle that had died of anthrax. As he looked through the lens of his microscope at the specimens, he spied rod-shaped bacilli, which he began to suspect were the culprits. (Bacilli are bacteria—see glossary.) He set out to track the entire life cycle of the bacillus, infecting mice with the disease and passing anthrax bacilli (from the blood of an infected animal) from one mouse to another through 20 generations. Shortly before, the German botanist Ferdinand Cohn had observed that a bacillus forms spores and had recognized their resistance to very high temperatures, and Koch now found that the anthrax bacillus forms spores that, as Pasteur also concluded, can survive in the earth for years. Koch succeeded in working out the entire life cycle of the bacillus, and Cohn enthusiastically sponsored the publication of his results.

Among Koch's many contributions to the growing understanding of the causes of infectious disease, he established rules for identifying the agent causing a disease. The researcher, he said, must locate the suspected microorganism in the diseased animal and then grow a pure strain of it in a culture. Then the cultured agent must cause the disease when introduced into a healthy animal. And the researcher must find the same kind of bacteria in the newly diseased animal as found in the original.

From his triumph with anthrax, Koch went on to establish improved methods for growing pure cultures of bacteria (using a gelatinous medium called agar-agar, composed from seaweed, and a culture dish invented by his assistant, Julius Richard Petri). He also succeeded in discovering the cholera bacillus and, in 1882, the tubercle bacillus, the cause of tuberculosis. His search for a cure for tuberculosis, unfortunately, met with frustration, even though at one point he thought he had found it.

In 1905, Koch received the Nobel Prize in medicine or physiology, primarily for his work on the causes of tuberculosis.

physician in Germany, had found a germ he thought was the cause of anthrax (see box). Using a microscope, Pasteur confirmed Koch's findings, and also found that the germ's spores were highly resistant to heat and could live in the ground over very long periods of time. An entire herd could become infected just by walking over the ground. Kill the affected animals, he said, burn them and bury them deep beneath the ground. He also came up with a way to inoculate against anthrax. There was no "milder" form of the disease to use as a vaccination, as Edward Jenner had done when he had used cowpox to protect against the virulent smallpox virus. So Pasteur heated some of the anthrax germs, reducing their virulence (potential for infection), and vaccinated sheep with them, leaving some sheep unvaccinated. The unvaccinated sheep all developed anthrax and died; the vaccinated sheep did not. Pasteur had developed an anthrax vaccine.

Pasteur used similar methods in developing vaccines against rabies and chicken cholera.

Pasteur died in 1895, having achieved enormous stature in the eyes of the world. He had won innumerable battles, most of them vast, with innovative techniques and unflappable genius. Unquestionably he was a hero in his own time and remains so to this day.

EPILOGUE

*Fortunately science, like that nature to which it belongs, is
neither limited by time nor by space. It belongs to the world,
and is of no country and no age. The more we know,
the more we feel our ignorance; the more we feel
how much remains unknown. . . .*

—Humphry Davy
November 30, 1825

The 19th century was a giant era in the history of science, a time when major discoveries opened up new worlds to explore—the worlds of atomic theory and dozens of new elements; of thermodynamics, electricity and electromagnetism; of diverse, evolving species and dinosaur bones; of plant and animal cells and of tiny, pathological organisms. New tools and methods such as electrolysis and spectroscopy provided keys to the elements, the stars and the universe. Scientists learned from each other (as Faraday did from Davy and Maxwell did from Faraday), vied with each other for priority (as Davy did with nearly everyone), were gracious with each other (in the manner of Darwin and Wallace) and argued issues assiduously (as Huxley did with Lyell and others). It was a time when science at last came into its own as a profession.

But by the end of the century, the very fabric of science was on the brink of change. What for Dalton, Faraday, Le Verrier, Maxwell and Helmholtz had seemed a noble pursuit of absolutes, of ultimate truths, was about to shift mightily. In the 1890s, as a new generation was about to take the helm—the generation of Max Planck, Ernest Rutherford, Marie Curie, Wilhelm Roentgen, Niels Bohr and Albert Einstein—certain absolutes seemed firm: Newtonian mechanics with its three-dimensional space and linear time, the laws of thermodynamics, Maxwell's electromagnetic waves in an all-pervading ether. But the dawn of the 20th century would bring extraordinary, mind-boggling change to all of these. So far, only the tip of the iceberg of science had been touched.

When Max Planck began the study of physics in the late 1870s, one of his teachers advised him not to go into the field—a few loose ends remained, he was told, but on the whole, all the major discoveries had already been made. But, as it has turned out, Humphry Davy's words spoken in 1825 held no less true at the end of the century than at the beginning, and they still hold true to this day.

A P P E N D I X   1

THE SCIENTIFIC METHOD

*. . . Our eyes once opened, . . . we can never go back to the old
outlook. . . . But in each revolution of scientific thought new
words are set to the old music, and that which has gone before
is not destroyed but refocused.*
—A. S. Eddington

What is science? How is it different from other ways of thinking? And
what are scientists like? How do they think and what do they mean
when they talk about "doing science"?

Science isn't just test tubes or strange apparatus. And it's not just frog
dissections or names of plant species. Science is a way of thinking, a vital,
ever-growing way of looking at the world. It is a way of discovering how the

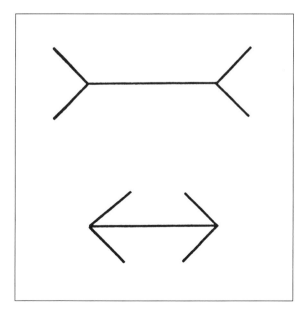

*Looks can be deceiving:
These two lines are the
same length.*

world works—a very particular way that uses a set of rules devised by scientists to help them also discover their own mistakes.

Everyone knows how easy it is to make a mistake about the things you see or hear or perceive in any way. If you don't believe it, look at the two horizontal lines on the previous page. One looks like a two-way arrow; the other has the arrow heads inverted. Which one do you think is longer (not including the "arrow heads")? Now measure them both. Right; they are exactly the same length. Because it's so easy to go wrong in making observations and drawing conclusions, people developed a system, a "scientific method," for asking "How can I be sure?" If you actually took the time to measure the two lines in our example, instead of just taking our word that both lines are the same length, then you were thinking like a scientist. You were testing your own observation. You were testing the information that we had given you that both lines "are exactly the same length." And, you were employing one of the strongest tools of science to do your test—you were quantifying, or measuring, the lines.

Some 2,400 years ago, the Greek philosopher Aristotle told the world that when two objects of different weights were dropped from a height, the heaviest would hit the ground first. It was a common-sense argument. After all, anyone who wanted to try a test could make an "observation" and see that if you dropped a leaf and a stone together the stone would land first. Try it yourself with a sheet of notebook paper and a paperweight in your living room. Not many Greeks tried such a test, though. Why bother when the answer was already known? And, being philosophers who believed in the power of the human mind to simply "reason" such things out without having to resort to "tests," they considered such an activity to be intellectually and socially unacceptable.

Centuries later, Galileo Galilei, a brilliant Italian who liked to figure things out for himself, did run some tests. Galileo, like today's scientists, wasn't content merely to observe the objects falling. Using two balls of different weight, a time-keeping device, and an inclined plane, or ramp, he allowed the balls to roll down the ramp and carefully *measured* the time it took. And, he did this not once, but many times, inclining planes at many different angles. His results, which still offend the common sense of many people today, demonstrated that, if you discount air resistance, all objects would hit the ground at the same time. In a perfect vacuum (which couldn't be created in Galileo's time), all objects released at the same time from the same height would fall at the same rate! You can run a rough test of this yourself (although it is by no means a really accurate experiment), by crumpling the notebook paper into a ball and then dropping it at the same time as the paperweight.

Galileo's experiments (which he carefully recorded step by step) and his conclusions based on these experiments demonstrate another important attribute of science. Anyone who wanted to could duplicate the experiments and either verify his results, or, by looking for flaws or errors in the experiments, prove him partially or wholly incorrect. No one ever proved Galileo wrong. And years later, when it was possible to create a vacuum (even though his experiments had been accurate enough to win everybody over long before that), his conclusions passed the test.

Galileo had not only shown that Aristotle had been wrong. He demonstrated how, by observation, experiment and quantification, Aristotle, if he had so wished, might have proven himself wrong—and thus changed his own opinion! Above all else the scientific way of thinking is a way to keep yourself from fooling yourself—or, from letting nature (or others) fool you.

Of course, science is more than observation, experimentation and presentation of results. No one today can read a newspaper or a magazine without becoming quickly aware of the fact that science is always bubbling with "theories." "ASTRONOMER AT X OBSERVATORY HAS FOUND STARTLING NEW EVIDENCE THAT THROWS INTO QUESTION EINSTEIN'S THEORY OF RELATIVITY," says a magazine. "SCHOOL SYSTEM IN THE STATE OF Y CONDEMNS BOOKS THAT UNQUESTIONINGLY ACCEPT DARWIN'S THEORY OF EVOLUTION," proclaims a newspaper. "BIZARRE NEW RESULTS IN QUANTUM THEORY SAY THAT YOU MAY NOT EXIST!" shouts another paper. What is this thing called "theory"?

Few scientists pretend any more that they make use of the completely "detached" and objective "scientific method" proposed by the philosopher Francis Bacon and others at the dawn of the scientific revolution in the 17th century. This "method," in its simplest form, proposed that in attempting to answer the questions put forward by nature, the investigator into nature's secrets must objectively and without preformed opinions observe, experiment and gather data about the phenomena. After Isaac Newton demonstrated the universal law of gravity, some curious thinkers suggested that he might have an idea *what gravity was*. But he did not see such speculation as part of his role as a scientist. "I make no hypotheses," he asserted firmly. Historians have noted that Newton apparently did have a couple of ideas, or "hypotheses," as to the possible nature of gravity, but for the most part he kept these private. As far as Newton was concerned there had already been enough hypothesizing and too little attention paid to the careful gathering of testable facts and figures.

Today, though, we know that scientists don't always follow along the simple and neat pathways laid out by the trail guide called the "scientific method." Sometimes, either before or after experiments, a scientist will get an idea or a hunch (that is, a somewhat less than well-thought-out hypothesis) that suggests a new approach or a different way of looking at a problem. Then he or she will run experiments and gather data to attempt to prove or disprove this hypothesis. Sometimes the word *hypothesis* is used more loosely in everyday conversation, but for a hypothesis to be valid scientifically it must have built within it some way that it can be proven wrong, if, in fact, it is wrong. That is, it must be falsifiable.

Not all scientists actually run experiments themselves. Most theoreticians, for instance, map out their arguments mathematically. But hypotheses, to be taken seriously by the scientific community, must always carry with them the seeds of falsifiabililty by experiment and observation.

To become a theory, a hypothesis has to pass several tests. It has to hold up under experiments, not only to the scientist conducting the experiments or making the observations, but to others performing other experiments and observations as well. Then, when thoroughly reinforced by continual testing and appraising, the hypothesis may become known to the scientific or popular world as a "theory."

It's important to remember that even a theory is also subject to falsification or correction. A good theory, for instance, will make "predictions"—events that its testers can look for as a further test of its validity. By the time most well-known theories such as Einstein's theory of relativity or Darwin's theory of evolution reach the textbook stage, they have survived the gamut of verification to the extent that they have become productive working tools for other scientists. But in science, no theory can be accepted as completely "proven"; it must always remain open to further tests and scrutiny as new facts or observations emerge. It is this insistently self-correcting nature of science that makes it both the most demanding and the most productive of humankind's attempts to understand the workings of nature. This kind of critical thinking is the key element of doing science.

The cartoon-version scientist portrayed as a bespectacled, rigid man in a white coat, certain of his own infallibility, couldn't be farther from reality. Scientists, both men and women, are as human as the rest of us, and they come in all races, sizes and appearances, with and without eyeglasses. As a group, because their methodology focuses so specifically on fallibility and critical thinking, they are probably even more aware than the rest of us of how easy it is to be wrong. But they like being right whenever possible and they like working toward finding the right answers to questions. That's usually why they have become scientists.

A P P E N D I X   2

ELEMENTS AND THEIR SYMBOLS

	Atomic Number		Atomic Number		Atomic Number
Ac, actinium	89	Cm, curium	96	Hg, mercury	80
Ag, silver	47	Co, cobalt	27	Ho, holmium	67
Al, aluminum	13	Cr, chromium	24	I, iodine	53
Am, americium	95	Cs, cesium	55	In, indium	49
Ar, argon	18	Cu, copper	29	Ir, iridium	77
As, arsenic	33	Dy, dysprosium	66	K, potassium	19
At, astatine	85	Er, erbium	68	Kr, krypton	36
Au, gold	79	Es, einsteinium	99	Ku, kurchatovium*	104
B, boron	5	Eu, europium	63	La, lanthanum	57
Ba, barium	56	F, fluorine	9	Li, lithium	3
Be, beryllium	4	Fe, iron	26	Lr, lawrencium	103
Bi, bismuth	83	Fm, fermium	100	Lu, lutetium	71
Bk, berkelium	97	Fr, francium	87	Md, mendelevium	101
Br, bromine	35	Ga, gallium	31	Mg, magnesium	12
C, carbon	6	Gd, gadolinium	64	Mn, manganese	25
Ca, calcium	20	Ge, germanium	32	Mo, molybdenum	42
Cd, cadmium	48	H, hydrogen	1	N, nitrogen	7
Ce, cerium	58	Ha, hahnium*	105	Na, sodium	11
Cf, californium	98	He, helium	2	Nb, niobium	41
Cl, chlorine	17	Hf, hafnium	72	Nd, neodymium	60

* name not officially approved

	Atomic Number		Atomic Number		Atomic Number
Ne, neon	10	Ra, radium	88	Tb, terbium	65
Ni, nickel	28	Rb, rubidium	37	Tc, technetium	43
No, nobelium	102	Re, rhenium	75	Te, tellurium	52
Np, neptunium	93	Rh, rhodium	45	Th, thorium	90
O, oxygen	8	Rn, radon	86	Ti, titanium	22
Os, osmium	76	Ru, ruthenium	44	Tl, thallium	81
P, phosphorus	15	S, sulfur	16	Tm, thulium	69
Pa, protactinium	91	Sb, antimony	51	U, uranium	92
Pb, lead	82	Sc, scandium	21	V, vanadium	23
Pd, palladium	46	Se, selenium	34	W, tungsten	74
Pm, promethium	61	Si, silicon	14	Xe, xenon	54
Po, polonium	84	Sm, samarium	62	Y, yttrium	39
Pr, praseodymium	59	Sn, tin	50	Yb, ytterbium	70
Pt, platinum	78	Sr, strontium	38	Zn, zinc	30
Pu, plutonium	94	Ta, tantalum	73	Zr, zirconium	40

CHRONOLOGY

1800	Alessandro Volta announces his invention (actually invented the previous year) of an electric battery.
1800	William Herschel detects infrared light.
1801	Johann Ritter discovers ultraviolet light.
1801	Giuseppe Piazzi discovers Ceres.
1801	Georges Cuvier identifies 23 species of extinct animals, adding fuel to the argument about the permanency of species.
1802	Thomas Young develops the wave theory of light.
1803	After studying meteorites found in France, Jean Baptiste Biot argues that the strange "stones" did not originate on Earth.
1804	Richard Trevithick builds a locomotive that pulls five loaded coaches along a track for nine and a half miles.
1804	Napoleon I is crowned emperor of France.
1805	Joseph Marie Jacquard develops the Jacquard loom. To control the operation of the loom Jacquard uses a system of punched cards, an idea that will later be incorporated in the design of early computers.
1807	Robert Fulton builds the steamship *Clermont*.
1807	In England, the Geological Society of London becomes the world's first official institution devoted solely to the study of geology.
1807	Coal-gas lighting begins to illuminate the streets of London.
1808	Humphry Davy develops the first electric-powered lamp.
1808	*The New System of Chemical Philosophy* by John Dalton revolutionizes chemistry.
1809	Jean-Baptiste de Monet de Lamarck publishes *Zoological Philosophy*.
1810	Davy shows that chlorine is an element.
1811	Amedeo Avogadro proposes what is now known as Avogadro's hypothesis.
1811	Herschel develops his theory of the development of stars and nebulas.
1812	Cuvier discovers the fossil of a pterodactyl.
1812	Pierre Simon de Laplace suggests that the universe can be viewed as a vast machine and that if the mass, position and velocity of

every particle could be known, the entire past and future of the universe could be calculated.

1814   Joseph von Fraunhofer rediscovers and charts solar spectral lines.

1814   George Stephenson introduces his first steam locomotive.

1815   Davy invents the safety lamp to be used by coal miners.

1815   William Prout suggests that hydrogen is the fundamental atom and that all other atoms are built up from different numbers of the hydrogen atom. Prout makes his first speculation anonymously since he himself thinks that the idea may be too extravagant.

1815   Napoleon is defeated at Waterloo.

1815   John Loudan McAdam constructs the first truly paved road.

1816   Augustin Fresnel demonstrates the wave nature of light.

1817   Fresnel and Thomas Young demonstrate that light waves must be transverse vibrations.

1818   Fresnel publishes "Memoir on the Diffraction of Light."

1818   Johann Franz Encke discovers what is now known as Encke's Comet.

1818   Jöns Jacob Berzelius publishes his table of atomic weights.

1819   Hans Christian Ørsted discovers that magnetism and electricity are two different manifestations of the same force. He does not publish this theory, though, until 1820.

1819   Pierre Louis Dulong and Alexis Thérèse Petit show that the specific heat of an element is inversely proportional to its atomic weight.

1820   The Royal Astronomical Society is founded in London.

1821   Michael Faraday demonstrates that electrical forces can produce motion (the first electric motor).

1822   Jean-Baptiste Joseph Fourier demonstrates Fourier's theorem and publishes his *Théorie analytique de la chaleur* ("Analytical Theory of Heat").

1822   Charles Babbage proposes the first modern computer but does not have the necessary modern materials with which to build it.

1823   John Herschel suggests that the so-called Fraunhofer lines might indicate the presence of metals in the sun.

1824   Nicolas Léonard Sadi Carnot publishes "On the Motive Power of Fire."

1824   The first school for science and engineering opens in the United States. It will eventually become the Rensselaer Polytechnic Institute.

1825   George Stephenson builds an improved steam locomotive.

1827   Georg Simon Ohm proposes what is now called Ohm's Law.

1827   Robert Brown reports his observation of the phenomenon now called "Brownian motion," which years later would help scientists to offer further proof of the existence of atoms.

1828   Friedrich Wöhler synthesizes urea, offering proof against the vitalist view that only living tissue could create organic molecules.

1830   Charles Lyell publishes the first volume of *The Principles of Geology*, offering evidence for the uniformitarian theory of the Earth's geological history.

1831   Charles Darwin begins his five-year voyage aboard the *Beagle*. With him is the first volume of Lyell's book.

1831   Michael Faraday discovers electromagnetic induction and devises the first electric generator. The discovery is made almost at the same time by the American scientist Joseph Henry.

1832   Faraday announces what are now called the laws of electrolysis.

1833   At a meeting of the British Association for the Advancement of Science, William Whewell proposes the term *scientist*.

1834   Cyrus Hall McCormick patents the McCormick Reaper.

1835   Gaspard Gustave Coriolis announces the "Coriolis Effect."

1837   Darwin begins to put together his theory of evolution but does not publish.

1838   Friedrich Bessel announces first precise measurement (using parallax) of the distance to a star.

1838   Matthias Jakob Schleiden announces his theory that all living plant tissue is made up of cells.

1839   Theodor Schwann extends Schleiden's cell theory to animals as well.

1839   Louis Jacques Mandé Daguerre invents the daguerreotype, an early type of photography.

1840   John William Draper takes the first photograph of the Moon.

1840   Germain Henri Hess founds the science of thermochemistry.

1842   Julius Robert Mayer becomes the first scientist to state both the mechanical equivalent of heat and the law of the conservation of energy, though not as well as Joule and Helmholtz, respectively.

1842   Christian Johann Doppler points out the phenomena of sound and other emissions from moving sources now known as the "Doppler shift."

1843   Samuel Heinrich Schwabe announces his discovery of the cyclic action of sunspots. The discovery begins early work in solar physics and astrophysics.

1844   Samuel F. B. Morse patents his design for the telegraph.

1846   The planet Neptune is discovered by Urbain Jean Joseph Le Verrier.

1846   The Smithsonian Institution is founded in America.

1847	Hermann von Helmholtz proposes the law of the conservation of energy (first law of thermodynamics).
1848	Lord Rosse discovers the Crab Nebula.
1848	*The Communist Manifesto* by Karl Marx and Friedrich Engels is published; revolutions sweep Europe.
1849	Jean Léon Foucault detects spectral emission lines.
1850	James Prescott Joule publishes his final figure for the mechanical equivalent of heat.
1850	Rudolf Clausius becomes the first scientist to clearly state the second law of thermodynamics.
1850	W. C. Bond of Harvard University makes the first astronomical photograph.
1851	William Thomson (later to be known as Lord Kelvin) proposes the concept of absolute zero.
1851	The Great International Exhibition opens in London, promoting the application of science to technology.
1851	Foucault demonstrates the rotation of the Earth.
1852	Edward Frankland publishes his theory of what came to be called "valence" in chemistry.
1852	Elisha Graves Otis builds the first safety elevator.
1853	Helmholtz proposes ideas about the ages of the Sun and the Earth based on his studies of the conservation of energy.
1855	Robert Bunsen and Gustav Kirchhoff begin work on the basics of spectral analysis.
1856	First skeleton of what we now call Neanderthals is found in a cave in Neander Valley, near Dusseldorf, Germany.
1856	Henry Bessemer develops the "Bessemer Process" and begins to build new "blast furnaces" that will open up the era of inexpensive steel.
1856	Louis Pasteur develops what is now known as "pasteurization."
1858	Darwin and Alfred Wallace announce their theory of evolution by natural selection to the Linnean Society.
1858	Rudolf Virchow publishes *Cellular Pathology*.
1858	The first transatlantic telegraph cable is laid.
1859	Darwin's *On the Origin of Species* is published.
1859	Edwin Drake drills first oil well near Titusville, Pennsylvania.
1859	Gaston Plante develops first storage battery.
1859	Kirchhoff and Bunsen announce their ideas on spectral lines.
c. 1860	James Clerk Maxwell develops the kinetic theory of gases. Ludwig Boltzmann also develops the theory independently and publishes his papers in the 1870s.
1860	Pasteur adds final argument against the long-held but increasingly shaky theory of spontaneous generation.

1860   Pierre Berthelot's work in synthesizing such organic molecules as methyl alcohol, ethyl alcohol and methane adds further proof of the ability of chemists to synthesize organic molecules from the elements, thus adding another blow to vitalist theories.

1860   Stanislao Cannizzaro wins chemists over to Avogadro's hypothesis with his speech and pamphlet at the First International Chemical Conference at Karlsruhe.

1860   Kirchhoff suggests that a body that absorbed all light and reflected none (called a black body) would, when heated, emit all wavelengths of light. This simple idea leads to questions that will help open up the next great revolution in physics in the early days of the 20th century.

1861   San Francisco and New York City are connected by a telegraph line.

1862   Foucault offers a new estimate for the velocity of light.

1862   Pasteur publishes his evidence for his germ theory of disease.

1863   The National Academy of Sciences is founded in the United States.

1863   William Huggins, after studying the spectra of some bright stars, announces that their spectral lines are those of familiar elements.

1864   Huggins, making the first spectrum analysis of a nebula, proposes that it is composed of gas.

1865   Gregor Mendel's theory of dominant and recessive genes is published in a little-known journal and goes unnoticed until early 1900s.

1865   Clausius coins the term *entropy*, describing the degradation of energy in a closed system.

1867   Marx publishes *Das Kapital.*

1867   The first truly functional typewriter is developed by Christopher Sholes.

1868   Helium is discovered by Pierre Jules César Janssen while he is studying the spectral lines of the sun.

1868   The first known Cro-Magnon skeletons are discovered in a cave in France.

1869   The Suez Canal opens.

1869   Dmitry Mendeleyev publishes his "periodic table of the elements."

1869   The final spike goes in place connecting the first transcontinental railway line in America.

1870   The businessman and amateur archeologist Heinrich Schliemann discovers the ancient city of Troy, uncovering vast amounts of gold and valuable objects and making the study of archeology a part of the popular consciousness.

1871 Darwin publishes *The Descent of Man.*

1872 Henry Draper is the first to photograph the spectra of a star.

1873 Maxwell publishes his theory of electromagnetism.

1875 Sir William Crookes develops the radiometer.

1876 Alexander Graham Bell patents the telephone.

1876 Nikolaus August Otto develops the four-cycle engine, the basis for today's internal-combustion engines.

1876 Josiah Willard Gibbs applies the theory of thermodynamics to chemical change.

1876 Repeating the work done by Julius Plücker almost two decades before, Eugen Goldstein describes the phenomena of cathode rays and is the first to use the term.

1877 Thomas Alva Edison invents the phonograph.

1879 Albert Michelson determines the velocity of light.

1879 Edison invents the incandescent electric light.

1880 Herman Hollerith develops the first electromechanical calculator. It is the next step toward today's modern computers.

1882 In America the Pearl Street Power Station brings electric lighting to New York City.

1884 Ottmar Mergenthaler patents the Linotype machine.

1885 Carl Friedrich Benz develops the first working automobile with a gasoline-burning internal-combustion engine.

1887 Albert Michelson and Edward Morley attempt to measure the changes in the velocity of light produced by the motion of Earth through space. Their failure to find any changes leads to the abandonment of belief in the ether and helps open the doorway to 20th-century physics.

1887 Heinrich Rudolph Hertz makes the first observation of the photoelectric effect; his observations will prove of great importance to the physics of the coming century.

1888 Hertz produces and detects radio waves and gives experimental evidence for the electromagnetic theory of Maxwell.

1889 The Eiffel Tower is finished in Paris. At the time it is the world's tallest human-made free-standing structure.

1889 Edward Charles Pickering makes the first observations of spectroscopic binary stars.

1890s Edison, borrowing on the ideas of others, develops the first successful motion pictures.

1894 The discovery of "Java Man" is announced by Marie Eugène Dubois.

1895 Sir William Ramsay discovers the element helium on Earth and finds that it would lie between hydrogen and lithium in the periodic table.

1895 Edward Emerson Barnard photographs the Milky Way.

1895 Wilhelm Konrad Röntgen discovers X rays.

1896 Antoine Henri Becquerel discovers natural radioactivity.

1897 J. J. Thomson discovers the electron.

1898 Marie and Pierre Curie isolate the radioactive elements of radium and plutonium.

GLOSSARY

atom the basic unit of an element; originally, as conceived by Democritus, the smallest, indivisible particle of a substance

atomic weight a number representing the weight of one atom, usually expressed in relationship to an arbitrary standard. Today the isotope of carbon taken to have a standard weight of 12 is commonly used, but several other standards were also used in the 19th century

bacteria (singular: bacterium) a large and varied group of microorganisms, typically one-celled. They are found living in many different environments and they typically have no chlorophyll, multiply by simple division and can be seen only with a microscope. Some bacteria cause diseases such as pneumonia, tuberculosis and anthrax, while others are necessary for fermentation and other biological processes

caloric term used to describe heat as a fluid, common in the 18th century, used by Antoine Lavoisier, among others

calx 18th- and 19th-century term for oxidized metal (which we would call "rust"); originally, the Roman name for lime, which Humphry Davy discovered was oxidized calcium

cell the basic unit of all living organisms; a very small, complex unit of protoplasm, usually organized in large numbers

compound a substance formed when two or more atoms of different elements are chemically united

diffraction the breaking up of a ray of light into dark and light bands or into the colors of the spectrum

electrochemistry the science of the interaction of electricity and chemical reactions or changes

electrolysis chemical change produced in a substance by the use of electricity; used especially to analyze substances

electromagnetism magnetism caused by electrical charges in motion; the physics of electricity and magnetism

element a substance that cannot be broken down into simpler substances

120

entropy a measure of the degree of disorder in a substance or a system; entropy always increases and available energy diminishes in a closed system

evolution the theory that groups of organisms may change over a long period of time so that descendants differ from their ancestors

inert gas a gas that has few or no active properties and does not react with other substances

isomers chemical compounds that have the same composition and molecular weight—the same chemical formula—but differ in the arrangement of atoms within the molecules and have different chemical or physical properties

mixture a blend of substances not chemically bound to each other

molecule two or more atoms chemically bound together, the fundamental particle of a compound

natural selection the process by which those individuals of a species with characteristics that help them adapt to their specific environment tend to leave more progeny and transmit their characters, while those less able to adapt tend to leave fewer progeny or die out, so that there is a progressive tendency in the species to a greater degree of adaptation. The mechanism of evolution discovered by Charles Darwin

parallax an apparent shift in position of an object viewed from two different locations, for example, from Earth at two different positions on opposite sides of its orbit (achieved by observing the objects at six-month intervals). The greater the apparent change in position, or "annual parallax," the nearer the star

pneumatic having to do with gases

polarize to produce polarization, that is, the condition of light or radiated energy in which the orientation of wave vibrations is confined to one plane or one direction only

refraction the bending of a ray or wave of light (or heat or sound) as it passes obliquely from one medium to another of different density, in which its speed is different, or through layers of different density in the same medium

thermodynamics the physics of the relationships between heat and other forms of energy

virus the smallest form of living organism, composed of nucleic acid and a protein coat; one of a group of ultramicroscopic or submicroscopic infective agents that cause various diseases in animals, such as measles, mumps and so on, or in plants, such as mosaic diseases. Viruses cannot replicate without the presence of living cells and are regarded both as living organisms and as complex proteins

vitalism the doctrine that the life in living organisms is caused and sustained by a vital force that is distinct from all physical and chemical forces. Kekulé's description of organic molecules as those containing carbon, with no reference to a life force or living matter, was a blow against the doctrine of vitalism

wavelength in a wave, the distance from one crest to the next, or from one trough to the next

F U R T H E R
R E A D I N G

ABOUT SCIENCE:

Cole, K. C. *Sympathetic Vibrations: Reflections on Physics as a Way of Life.* New York: William Morrow and Co., 1985. Well-written, lively, and completely intriguing look at physics presented in a thoughtful and insightful way by a writer who cares for her subject. The emphasis here is primarily modern physics and concentrates more on the ideas than the history.

Gardner, Martin. *Fads and Fallacies in the Name of Science.* New York: New American Library, 1986 (reprint of 1952 edition). A classic look at pseudoscience by the master debunker. Includes sections on pseudo-scientific beliefs in the 19th century.

Gonick, Larry, and Art Huffman. *The Cartoon Guide to Physics.* New York: Harper Perennial, 1991. Fun, but the whiz-bang approach sometimes zips by important points a little too fast.

Hann, Judith. *How Science Works.* Pleasantville, N.Y.: The Reader's Digest Association, Inc., 1991. Lively well-illustrated look at physics for young readers. Good, brief explanations of basic laws and short historical overviews accompany many easy experiments readers can perform.

Hazen, Robert M., and James Trefil. *Science Matters.* New York: Doubleday, 1991. A clear and readable overview of basic principles of science and how they apply to science in today's world.

Holzinger, Philip R. *The House of Science.* New York: John Wiley and Sons, 1990. Lively question-and-answer discussion of science for young adults. Includes activities and experiments.

Trefil, James. *1001 Things Everyone Should Know about Science.* New York: Doubleday, 1992. The subtitle, *Achieving Scientific Literacy*, tells all. Well done for the average reader but includes little history.

ABOUT THE HISTORY OF
NINETEENTH-CENTURY SCIENCE:

Asimov, Isaac. *Asimov's Biographical Encyclopedia of Science and Technology.* Second Revised Edition. Garden City, N.Y.: Doubleday and Company, 1982. This book's unusual nonalphabetical, chronological entry system takes some getting used to, but overall its lively and typically opinionated Asimov approach makes for fascinating reading as well as basic fact gathering.

————. *Asimov's Chronology of Science and Discovery.* New York: Harper and Row, 1989. A lively chronological view of science, year by year. Written with Asimov's usual verve. Good for fact-checking and browsing.

Boorstin, Daniel J. *The Discoverers.* New York: Random House, 1983. Its size may be intimidating—over 700 pages—but this is a wonderfully lively, thoughtful and absorbing look at the history of humankind's search to know itself and nature. Aimed at the general reader.

Brooke, John Hedley. *Science and Religion: Some Historical Perspectives.* Cambridge: Cambridge University Press, 1991. Well-balanced and thoughtful look at the sometimes troubled relationship between science and religion.

Hellemans, Alexander, and Bryan Bunch. *The Timetables of Science.* New York: Simon and Schuster, 1988. Easy-to-read chronology of the history of science. Good for fact-finding or simply browsing. And the "overviews" are nicely done and add some colorful context.

Jones, Bessie Zaban, ed. *The Golden Age of Science: Thirty Portraits of the Giants of 19th-Century Science by Their Scientific Contemporaries.* New York: Simon and Schuster, 1966.

Knight, David. *The Age of Science: The Scientific World View of the 19th Century.* Oxford: Basil Blackwell Ltd., 1986.

Mackay, Charles. *Extraordinary Popular Delusions and the Madness of Crowds.* New York: Harmony Books, 1990. Reprint of the classic 1841 book on various money-making schemes and outlandish beliefs people have bought into. Provides a fascinating sense of the time in which it was written as well as ageless insights into human psychology and its struggles against logic. Includes original illustrations and a present-day foreword by business writer Andrew Tobias.

Mason, Stephen F. *A History of the Sciences.* Revised Edition. New York: Collier Books, 1962. Originally published in 1956, this book is older than Ronan's (see below) but is a solid standard history of science.

Meadows, Jack. *The Great Scientists.* Oxford: Oxford University Press, 1987. Easy to read, well presented and nicely illustrated.

Ronan, Colin. *The Atlas of Scientific Discovery*. New York: Crescent Books, 1983. A slim coffeetable-type book but a well-written and well-thought-out overview with illustrations.

———. *Science: Its History and Development Among the World's Cultures*. New York: Facts On File, 1982. A good, readable comprehensive overview of the history of science from the ancients to the present.

Shapiro, Gilbert. *A Skeleton in the Darkroom: Stories of Serendipity in Science*. San Francisco: Harper and Row, 1986. Well-told story, but it concentrates a little more heavily on the philosophical and sociological than on the scientific. Includes the colorful story about Ørsted's famous "accidental" discovery of the link between electricity and magnetism.

Williams, L. Pearce. *Album of Science: The Nineteenth Century*. New York: Charles Scribner's Sons, 1978.

ABOUT THE PHYSICAL SCIENCES:

Abbott, David, ed. *The Biographical Dictionary of Scientists: Physicists*. New York: Peter Bedrick Books, 1984. Like Abbott's dictionary of astronomers, a reliable reference, though somewhat tough going.

———. *The Biographical Dictionary of Scientists: Astronomers*. New York: Peter Bedrick Books, 1984. Short entries from A to Z, including an extensive glossary and line diagrams. Dry and a little difficult, but a good resource.

Asimov, Isaac. *Atom: Journey Across the Subatomic Cosmos*. New York: Truman Talley Books (Dutton), 1991. A clear, vivid, easy-to-read history of discoveries about the atom.

———. *Eyes on the Universe*. Boston: Houghton Mifflin Company, 1975. A good starting place for readers wanting to know more about the early history and development of telescopic instruments.

Boorse, Henry A., Lloyd Motz, and Jefferson Hane Weaver. *The Atomic Scientists: A Biographical History*. New York: John Wiley and Sons, 1989. Includes a brief but informative chapter on Dalton, Gay-Lussac and Avogadro. Concise and well written.

Buttmann, Gunther. *The Shadow of the Telescope: A Biography of John Herschel*. New York: Charles Scribner's Sons, 1970. Entertaining look at the life and time of William Herschel's son, John, a famous astronomer and personality in his own right.

Fleisher, Paul. *Secrets of the Universe: Discovering the Universal Laws*. New York: Atheneum, 1987. Well-written and engrossing look at physics aimed at the younger reader but enjoyable and informative for anyone interested in science. Includes excellent discussions of the work of Galileo, Newton, and others, as well as general laws of physics.

Gamow, George. *The Great Physicists From Galileo to Einstein.* New York: Dover Publications, Inc., 1988. The great Gamow takes a look at some major historical physicists and their work. As always with Gamow, readable and enlightening.

Harman, P. M. *Energy, Force, and Matter: The Conceptual Development of Nineteenth-Century Physics.* Cambridge: Cambridge University Press, 1982. A good look at the state of physics and the important ideas it was beginning to pursue in the 19th century. Sometimes a little rough going.

Hudson, John. *The History of Chemistry.* New York: Chapman and Hall, 1992. A highly readable account, including profiles on key scientists, photographs and diagrams.

McCormmach, Russell. *Night Thoughts of a Classical Physicist.* Cambridge: Harvard University Press, 1982. Set in Germany in the early days of the 20th century, a fascinating novel treating the mind and emotions of a physicist, trained in the approaches of 19th-century physics, who attempts to understand the disturbing changes in physics and the world.

Motz, Lloyd, and Jefferson Hane Weaver. *The Story of Physics.* New York: Avon Books, 1989. Occasionally rough going but well worth a try for a good, strong narrative history of the theories, ideas, experiments and people of physics.

Ruben, Samuel. *The Founders of Electrochemistry.* Philadelphia: Dorrance and Company, 1975. Readable account of some of the major figures in the early days of electrochemistry, with special emphasis on Davy and Faraday.

Spielberg, Nathan, and Bryon D. Anderson. *Seven Ideas That Shook the Universe.* New York: John Wiley and Sons, 1987. A fascinating book that centers on the drama of scientific discovery, including a thorough chapter on concepts of energy and entropy.

Walker, Jearl. *The Flying Circus of Physics with Answers.* New York: John Wiley and Sons, 1977. A now classic collection of problems and questions about physics in the everyday world. Not much history but much fun.

ON DARWIN AND THE THEORY OF EVOLUTION:

Bibby, Cyril. *T. H. Huxley: Scientist, Humanist and Educator.* New York: Horizon Press, 1960. A little dated and old-fashioned in approach but still an interesting look at the fascinating Huxley.

Bowlby, John. *Charles Darwin: A New Life.* New York: W. W. Norton Company, 1990. Off-the-beaten-path look at Darwin, concerned chiefly with his mysterious medical problems. Worth a look but only after pursuing many other better books on the subject.

Bowler, Peter J. *Evolution: The History of an Idea.* Revised Edition. Berkeley, Calif.: University of California Press, 1989. An excellent look at the history of evolutionary theory. Includes many of the subjects included in this book. Highly thoughtful and informative.

Burke, James. *The Day the Universe Changed.* Boston: Little Brown and Company, 1985. Includes a good chapter on Darwin and his predecessors in evolutionary theory as well as a good chapter on the important scientific investigations into the nature of light.

Burkhardt, F., and S. Smith, eds. *The Correspondence of Charles Darwin,* 7 vols. Cambridge: Cambridge University Press, 1985–91. The definitive edition of Darwin's letters.

Clark, Ronald W. *The Survival of Charles Darwin.* New York: Random House, 1984. A more or less traditional approach to the life of Darwin. Less "politically aware" than the more modern but occasionally annoying Desmond and Moore book (see below).

Darwin, Charles. *On the Origin of Species by Means of Natural Selection, or the Preservation of Favoured Races in the Struggle for Life. . .* London: Murray, 1859. Also available in many annotated and paperback editions.

———. *The Descent of Man and Selection in Relation to Sex,* 2 vols. London: Murray, 1871; 2nd ed. rev., 1874. Also available in annotated and paperback editions.

Desmond, Adrian. *Archetypes and Ancestors: Paleontology in Victorian London 1850–1875.* Chicago: University of Chicago Press, 1982.

Desmond, Adrian, and James Moore. *Darwin.* New York: Warner Books, 1991. Desmond and Moore make use of much new material and offer a modern and updated look at Charles Darwin, his life and work. Emphasizes the social and political background of Darwin's thought and sometimes tries to rewrite history in modern terms. Still, it's the best modern book on Darwin.

Edey, Maitland, and Donald C. Johanson. *Blueprints: Solving the Mystery of Evolution.* Boston: Little, Brown and Co., 1989. Engrossingly told story coauthored by the paleoanthropologist, Johanson, the discoverer of the famous "Lucy."

Eldredge, Niels. *The Monkey Business: A Scientist Looks at Creationism.* New York: Washington Square Press, 1982. A top-drawer evolutionary scientist answers the creationists' attack on evolutionary theory.

Gould, Stephen Jay. *Ever Since Darwin: Reflections in Natural History.* New York: W. W. Norton and Company, 1977. Includes an enlightening discussion on early ideas about preformation.

Hays, H. R. *Birds, Beasts, and Men: A Humanist History of Zoology.* New York: G. P. Putnam's Sons, 1972. Readable and well-organized narrative but old and may be hard to find.

Irvine, William. *Apes, Angels, and Victorians: The Story of Darwin, Huxley, and Evolution.* New York: McGraw-Hill, 1955. A colorful and readable look at Darwin and Huxley, filled with the atmosphere and scientific controversies of Victorian England.

Jastrow, Robert, and Kenneth Korey. *The Essential Darwin.* Boston: Little, Brown and Company, 1984. Readings from Darwin with editorial comments and explanations by Korey.

Lanham, Url. *The Bone Hunters.* New York: Columbia University Press, 1973. A very readable account of the work, adventures and rivalry of Marsh and Cope.

Lewin, Roger. *Thread of Life: The Smithsonian Looks at Evolution.* Washington, D.C.: Smithsonian Books, 1982. Well-illustrated and well-produced look at the science and history of evolution.

Magner, Lois N. *A History of the Life Sciences.* New York: Marcel Dekker, Inc., 1979. A good, readable overview but marred somewhat by awkward organization.

Mayr, Ernst. *The Growth of Biological Thought: Diversity, Evolution, and Inheritance.* Cambridge, Mass.: Harvard University Press, 1982. Impressive if sometimes heavy-going look at the history of evolutionary theory by one of the world's leading experts.

———. *One Long Argument: Charles Darwin and the Genesis of Modern Evolutionary Thought.* Cambridge, Mass.: Harvard University Press, 1991.

McGowan, Chris. *In The Beginning: A Scientist Shows Why the Creationists Are Wrong.* Buffalo: Prometheus Books, 1984. For students or anyone wishing to understand more about the still heated evolution versus creationist debate. McGowan's book offers a reasoned attack against the creationist arguments.

Miller, Jonathan, and Boran Van Loon. *Darwin for Beginners.* New York: Pantheon Books, 1982. A nifty cartoon presentation that packs a lot of information into the format. A good overview for anyone wanting a quick understanding of Darwin and his theory, but it should be accompanied by some other books on the subject.

Milner, Richard. *The Encyclopedia of Evolution: Humanity's Search for Its Origins.* New York: Facts On File, 1990. Totally engrossing for hunting facts or browsing, one of the few books of its kind that is also completely rewarding, entertaining, and informative as just plain reading. Quirky and fascinating and chock-full of just about everything you wanted to know about evolution but were afraid to ask.

Reader, John. *Missing Links: The Hunt for Earliest Man.* Boston: Little, Brown and Company, 1981. The hunt for the so-called "missing link" between ape and human that began with the discovery of the first Neanderthal skeleton in 1856.

Sears, Paul B. *Charles Darwin: The Naturalist as a Cultural Force.* New York: Charles Scribner's Sons, 1950. Brief and interesting look at Darwin and his influence.

Skelton, Renne. *Charles Darwin and the Theory of Natural Selection.* New York: Barrons, 1987.

Stein, Sara. *The Evolution Book.* New York: Workman Publishing, 1986. Written for younger people, but fun and instructive for all, a well-done explanation of evolution, with many examples and projects.

Trinkaus, Erik, and Pat Shipman. *The Neandertals: Changing the Image of Mankind.* New York: Alfred A. Knopf, 1993. Just about the latest and most up-to-date information on the new light being shed on the once poorly understood Neanderthals. A good, solid book, nicely re-searched, but a little difficult to read.

ON FARADAY AND MAXWELL:

May, Charles Paul. *James Clerk Maxwell and Electromagnetism.* New York: Franklin Watts, 1962.

Thomas, John Meurig. *Michael Faraday and the Royal Institution: The Genius of Man and Place.* Bristol: Adam Hilger (IOP Publishing, Ltd.), 1991. Written by the current director of the Royal Institution and the Davy Faraday Laboratory, this short, delightful study includes partial facsimiles of some of Faraday's manuscripts and photographs from the archives of the Royal Institution. Also covers the history of the Royal Institution.

Williams, L. Pearce. *Michael Faraday.* New York: Basic Books, Inc., 1964.

I N D E X

Italic numbers indicate illustrations.

A

absolute zero 116
acetic acid (vinegar) 22, 29
acquired characteristics 79
actinium 111
Adams, John Couch 63
agar-agar 102
Agassiz, Louis 70, 85
air 5, 7
alchemy xvi
alcohol 117
Aldebaran (star) 65
alizarin 31
aluminum 21, 111
amalgam 18
amber 45
American Association for the
  Advancement of Science xiii
americium 111
ammonia 22
ammonium cyanate 29
Ampère, André Marie xv, 48–49
amputation 101
Analytical Theory of Heat
  (Jean-Baptiste Joseph Fourier)
  36, 114
aniline 30
animal cells 95–96, 115
"animal electricity" 12, 14
anthrax 101–103
antimony 21, 112
antiseptic surgical technique 101
Apreece, Jane 18
Arabs 5
Arago, Dominique François 48

archaeology 117
argon 28, 111
Aristotle 5, 108–109
arsenic 111
aseptic surgical technique 101
Asimov, Isaac 98
astatine 111
astigmatism 56
astronomy 59–69, 116
atomic weight 8–10, 22, 24, 114,
  120
atoms 3–11, 33–34, 36, 114–115,
  120
Autobiography (Charles Darwin)
  78
automobiles 118
autonomic nervous system 94
Avogadro, Amedeo 9, 33, 113
Avogadro's hypothesis 9, 22, 113,
  117
Ayurvedic philosophy 4

B

Babbage, Charles 114
Bacon, Francis 81, 109
bacteria 102, 120
Baeyer, Adolf von 31
barium 18, 24, 64–65, 111
barnacles 82
Barnard, Edward Emerson *62, 63,*
  68, 119
Barnard, Sarah 51
baryta 18
batteries 12–17, *13,* 46, 113
Beagle (ship) 76, 115

Becquerel, Antoine Henri 119
Beddoes, Thomas 15
beer industry 99
Bell, Alexander Graham 118
Benz, Carl Friedrich 118
benzene ring 30–32, *33*
berkelium 111
Bernard, Claude 91–92, *92*, 93
Bernoulli, Daniel 36
Berthelot, Pierre 117
beryllium 111
Berzelius, Jöns Jacob 9, 18, 21–22,
  28–29, 114
Bessel, Friedrich 60–62, 115
Bessemer, Henry 116
Betelgeuse (star) 65
binary stars 118
biochemistry 92–93
Biot, Jean-Baptiste 29, 32, 113
bismuth 66, 111
Black, Joseph 6, 35, 38
black body 117
blast furnaces 116
blood flow 92
Boisbaudran, Paul Emile Lecoq de
  27–28
Boltzmann, Ludwig 42–43, 54,
  116
Bond, W. C. 116
bone hunters *80*, 80–81
boron 21, 111
Boyle, Robert 5–7
Brahe, Tycho xiv
brain 93–95
brine 12
British Association for the
  Advancement of Science xiii
bromine 21, 92, 111
Brown, Robert 95, 115
Brownian motion 115
brucine 92
Buffon, Georges Louis Leclerc de
  78

Bunsen, Robert Wilhelm 24, 27,
  64, 116
Bunsen burner 24, 27

C

cadmium 111
calcium 18, 21, 24, 65–66, 111
calculators 118
californium 111
caloric theory 16, 35–36, 120
calx 18, 120
Cannizzaro, Stanislao 22, 24, 117
carbolic acid (phenol) 101
carbon 10, 31–32, 111
carbon dioxide 5, 10
carbon monoxide 10, 93
Carnot, Nicolas Léonard Sadi
  39–41, 43, 114
cars 118
catastrophe theory 70
cathode rays 118
Cavendish, Henry 28
cell division 96
cells 95–98, 120
Cellular Pathology (Rudolf Carl
  Virchow) 97, 116
celluloid 31
Ceres (star) 113
cerium 111
cesium 27, 111
Chambers, Robert 82
Charles, Jacques 9
Charles's law 9
Chateaubriand, René de xvii
Chemical Conference, First
  International (Karlsruhe
  Conference) (1860) 22, 24
chemistry 3–34, 113, 116
childbed fever 100–101
China 4
chlorine 18, 21, 24, 111, 113
cholera 102–103

chromium 65, 111
Clark, Alvan 62–63
Clausius, Rudolf *40*, 40–41, 43, 116–117
Clermont (steamship) 113
closed system 42
coal-gas lighting 113
cobalt 111
codeine 92
Cohn, Ferdinand 102
Coleridge, Samuel Taylor 16
color blindness 11
colors 27
color theory 56
Commentaries (Luigi Galvani) 14
compasses 45
compounds 9–10, 120
computers 113–114, 118
conditioned reflex 95
conductors 14
conservation of energy, law of *See* thermodynamics, first law of
conservation of matter, law of 6
Cope, Edward Drinker 80–81
Copernicus xvii, 59
copper 14, 65, 111
cordite 30
Coriolis, Gaspard Gustave 115
coumarin 31
Crab Nebula 63, 116
Cro-Magnons 89, 117
Crombie, A. C. 38
Crookes, William 118
Curie, Marie 119
Curie, Pierre 119
curium 111
Cuvier, Georges 48, 70, 88, 113
Cygnus (constellation) 60

D

Daguerre, Louis Jacques Mandé 115

daguerreotype 115
Dalton, John *4*, 6–11
   atomic theory xv–xvii, 3, 8–11, 19, 33, 36
   early work 6–7
   publications 10, 113
   symbols used by *8*, 21
Darwin, Charles 75–90, 77, *82*, *84*, *86*
   Beagle voyage 72, 75–78, 115
   evolutionary theory xv–xvii, 69, 79–90, 110, 121
   publications xiii, 116, 118
Darwin, Erasmus (brother of Charles Darwin) 83
Darwin, Erasmus (grandfather of Charles Darwin) 75
Davy, Humphry 15–19, *17*, 56
   on atomic theory 3
   caloric theory opposed by 36
   elements discovered by 18–19, 21, 120
   Faraday as assistant to 48–49, 51
   on future xi
   inventions 113–114
   on science 105–106
definite proportions, law of 8
Democritus 3, 5–6, 8, 120
Descartes, René xiv, 36, 93–94
Descent of Man, and Selection in Relation to Sex, The (Charles Darwin) 87, 118
diffraction 56, 120
digestion 92–95
dinosaurs 80–81
disease *See* germ theory
divergence 82
Döbereiner, Johann Wolfgang 24
dogs 91, 94–95
dominant genes 117
Doppler, Christian Johann 115
Doppler shift 115

Drake, Edwin 116
Draper, Henry 118
Draper, John William 115
Dubois, Marie Eugène 118
Dulong, Pierre Louis 114
dyads 5
dyes 30–31
dynamite 30
dynamos *See* generators
dysprosium 111

E

Earth 69–72, 79, 115–116, 118
earth science *See* geology
Edison, Thomas Alva 52, *52*, 53, 118
egg cell (ovum) 96
Eiffel Tower 118
Einstein, Albert 55, 63, 110
einsteinium 111
Eldredge, Niles 90
electrical power stations 53
electric generators 50, 115
electricity 11–18 *See also*
   electromagnetism
electric light 53, 113, 118
electric motors 49, 114
electrochemistry 15–19, 120
electrolysis 16, 115, 120
electromagnetic fields 50, 54–55
electromagnetic induction 50, 115
electromagnetism 45–58, 114,
   118, 120
electrons 119
elements 3–19, 111–112, 120 *See*
   also periodic table
Elements of Agricultural Chemistry
   (Humphry Davy) 18
Elements of Chemical Philosophy
   (Humphry Davy) 18
elevators 116
embryos 78
empiricism 71

Encke, Johann Franz 114
Encke's comet 114
energy, law of conservation of *See*
   thermodynamics, first law of
Engels, Friedrich xii, 116
engines
   electric 49, 114
   internal-combustion 118
   steam 36, 38–39
entropy 42, 117, 121
epigenesis 78
erbium 111
ethyl alcohol 117
Euclid 66
europium 111
evolution 75, 78–90, 116, 121
explosives 30

F

Faraday, Michael xv, xvi, 18, 32,
   47, 47–55, *49*, 58, 114–115
fermentation 98–99
fermium 111
Ferrier, David 94
field theory 50–51
finches 77
FitzRoy, Robert 76–77
fluorine 111
fossils 79–81, 113 *See also*
   Cro-Magnons; Neanderthals
Foster, George Carey 25
Foucault, Jean Léon 116–117
four-cycle engine 118
Fourier, Jean-Baptiste Joseph 36,
   39, 43, 114
Fourier's theorem 36, 114
francium 111
Frankland, Edward 116
Franklin, Benjamin xiv–xv, 12,
   14–15, 46
Fraunhofer, Joseph von 26–27,
   60–63, 114

Fraunhofer lines 26–27, 64–67
Fresnel, Augustin 57–58, 114
Fritsch, Gustav 94
frogs 11–12, 14
Fulton, Robert 113

G

gadolinium 111
Galápagos Islands 77, 77–78
galaxies 63
Galileo Galilei xiv, 59, 108–109
Galle, Johann 63
gallium 28, 111
Galvani, Luigi 11–12, 14
galvanometer 14, 49
gases 5, 7, 9, 11, 15–16, 121 *See also* kinetic theory of gases
Gay-Lussac, Joseph 9, 22, 48
generators (dynamos) 50, 115
genes 117
Geological Evidence for the Antiquity of Man (Charles Lyell) 87
Geological Society of London 113
geology 69–72, *70*, 113
germanium 28, 111
germ theory 98–103, 117
Gibbs, Josiah Willard 118
Gilbert of Colchester, William 45, *46*
glass 45
glucose 93
glycogen 93
Goethe, Johann Wolfgang von xvi
gold 21, 111
Goldstein, Eugen 118
Gould, Stephen Jay 90
Graebe, Karl 31
Great Chain of Being 78
Great International Exhibition (London, 1851) 116
Greeks 4–5
Greig, Samuel 66

Grimaldi, Francesco 56
Guericke, Otto von 46

H

hafnium 111
hahnium 111
Hall, Asaph 62
Halsted, William Stewart 101
heat 35–43
  as fluid 16, 35–36, 120
  as motion 15, 36
heat engines 38
Hegel, Georg xvi
heliometer 61–62
helium 28, 111, 117–118
Helmholtz, Hermann von 38–39, *39*, 43, 55–56, 69, 116
Henry, Joseph 50, 115
Henslow, John Stevens 76–77
Herschel, John xiv, 66–68, *68*, 75, 114
Herschel, William 55, 59, 63, 68, 113
Hertz, Heinrich Rudolf 55–58, *58*, 118
Hess, Germain Henri 115
Higgins, William 10
Hindus 4
Hitzig, Julius 94
Hofmann, August Wilhelm von 31
Hollerith, Herman 118
holmium 111
Hooke, Robert 5, 95
Hooker, Joseph Dalton 82–84, 90
Huggins, William 65, *65*, 66–67, 117
human evolution 85–89
Humboldt, Alexander von 48, 76
Hutton, James 69, 72
Huxley, T. H. xvii, 85–86, *87*, 90
Hyatt, John Wesley 31
hydrogen 5, 9, 32, 111, 114

hypothesis 109–110

I

ice ages 70
"imponderables" xvi, 15, 35
incandescent electric light 53, 118
India 5
indigo 31
indium 111
induction *See* electromagnetic
  induction
Industrial Revolution xii–xiii
inert gases 121
infection *See* germ theory
infrared light 55, 113
inorganic chemistry 21–28
internal-combustion engines 118
iodine 21, 92, 111
iridium 111
iron 45, 65–66, 111
isomers 29–30, 121

J

Jacquard, Joseph Marie 113
Jacquard loom 113
Janssen, Pierre Jules César 69, 117
"Java Man" 118
Jenner, Edward 103
joule 38
Joule, James Prescott *37*, 37–38,
  41, 43, 58, 116
Jupiter (planet) 63

K

Kant, Immanuel 59
Karlsruhe Conference (First
  International Chemical
  Conference) (1860) 22, 24
Keats, John xvii
Kekulé von Stradonitz, Friedrich
  22, 30–32, 122

Kelvin, Lord (William Thomson)
  xi, xvi, 41, *41*, 43, 69, 116
Kepler, Johannes xiv, 59, 61
kinetic theory of gases 42–43,
  116
kinetic theory of heat 36
Kirchhoff, Gustav 27, 64–65, 67,
  116–117
Kleist, Ewald von 46
Koch, Robert 101–103
Kolbe, Adolph Wilhelm Hermann
  29
Kölliker, Rudolf Albert von 96
krypton 28, 111
kurchatovium 111

L

lactic acid 99
Lamarck, Jean-Baptiste de Monet
  de 79, 113
Lamartine, Alphonse de xvii
lanthanum 111
Laplace, Pierre Simon de 113
laughing gas (nitrous oxide) 15–16
Lavoisier, Antoine xvi, 6, 11, 15,
  48, 120
lawrencium 111
lead 112
Leucippus 5
Le Verrier, Urbain Jean Joseph
  63, *64*, 115
Leyden jar 14, 46
Lick Observatory 63
Liebig, Justus von 29, 36, 99
life sciences 75–103
light 55–58
  coal-gas 113
  electric 53, 118
  infrared 55, 113
  particle theory of 56–57
  polarized 29, 57
  speed of 117–118

ultraviolet 55, 113
   wave theory of 56–57, 113–114
lightning 14
lime (chemistry) 16, 18
Linotype machine 118
Lister, Joseph *100*, 101
lithium 21, 111
liver 93
locomotives 113–114
Lubbock, John 90
lutetium 111
Lyell, Charles *71*, 72
   burial site 90
   evolutionary theory and 69,
      75–76, 78, 83–85, 87
   geological theory 72, 115
   influences on 66

M

Mach, Ernst 33
Magendie, François 91–92
magnesia 16, 18
magnesium 18, 65–66, 111
magnetism *See* electromagnetism
Malthus, Thomas Robert 79
manganese 111
Marconi, Marchese Guglielmo 57
Mars (planet) 63
Marsh, Othniel C. 80–81
Marx, Karl xii, 116–117
material theory of heat *See* caloric
   theory
mathematical analysis 36
matter, law of conservation of 6
Maximilian I (Bavaria) 26
Maxwell, James Clerk 52–55, *54*
   electromagnetism 54–55,
      57–58, 118
   on Helmholtz's contribution 39
   kinetic theory of gases 42–43,
      54, 116
   on unification of science xvi

Maxwell's demon 42–43
Mayer, Julius Robert 38, 58,
   115
McAdam, John Loudan 114
McCormick, Cyrus Hall 115
mechanical equivalent of heat 43,
   115–116
Mendel, Gregor xv, 88–89, 96,
   117
mendelevium 24, 111
Mendeleyev, Dmitry *23*, 23–25,
   28, 33, 117
mercury 18, 111
Mercury (planet) 63
Mergenthaler, Ottmar 118
Messier, Charles 59
metal oxides 16
metals 12, 14
meteorites 113
methane 117
methyl alcohol 117
Michelson, Albert 118
Milky Way 68, 119
Mitchell, Maria 60, *61*
mixtures 7, 121
models, mechanical xvi
Mohr, Friedrich 36
molecules 9–10, 121
molybdenum 111
Moon 115
Morley, Edward 118
morphine 92
Morse, Samuel F. B. 115
motion pictures 53, 118
motor regions (brain) 94
motors *See* engines
multiple proportions, law of 10
Musschenbroek, Pieter van 46

N

Nägeli, Karl Wilhelm von 96
naphthalene 31

Napoleon Bonaparte xii, 12, 18, 56, 113–114
National Academy of Sciences 117
natural selection 79–80, 116, 121
Naturphilosophie (school of philosophy) xvi
Neanderthals 88–89, 116
nebulas 66–67, 113, 117 *See also* Crab Nebula
neodymium 111
neon 28, 112
Neptune (planet) 63, 115
neptunism (geology) 69
neptunium 112
nerves 92–93, 96
nervous system 94
Newlands, John Alexander Reina 25
New System of Chemical Philosophy, The (John Dalton) 7, 113
Newton, Isaac xiv–xv, 6, 26–27, 56, 61, 90, 109
nickel 65, 112
niobium 111
nitrocellulose 30–31
nitrogen 5, 7, 9, 21, 111
nitroglycerine 30
nitrous oxide (laughing gas) 15–16
nobelium 112
nucleus (cell) 95

O

octaves, law of 25
Ohm, Georg Simon 114
Ohm's Law 114
oil wells 116
On the Motive Power of Fire (Nicolas Léonard Sadi Carnot) 40
organic chemistry 28–32
Origin of Species (Charles Darwin) xiii, 85, 116

Orsted, Hans Christian xv, 45, 48–49, 114
osmium 112
Otis, Elisha Graves 116
Otto, Nikolaus August 118
ovum (egg cell) 96
Owens, Richard 85–86
oxygen 5, 7–10, 17, 21, 93, 112
oxymuriatic acid 18

P

paleontologists 80–81
palladium 112
pancreas 93
parallax 60, 115, 121
Parkes, Alexander 31
Parsons, William *See* Rosse, Lord
partial pressures, law of 7
particle theory of light 56–57
Pasteur, Louis 29, 32, 91, *98*, 98–103, 116–117
pasteurization 99, 116
Pavlov, Ivan 91, 94–95
peas 88
perfumes 31
periodic table 25–26, *26*, 28, 117
Perkin, William 30–32
Petit, Alexis Thérèse 114
Petri, Julius Richard 102
phenol (carbolic acid) 101
phlogiston xvi
phonograph 53, 118
phosphorus 112
photoelectric effect 118
photography 68, 115–116, 118
phrenology 94
physical sciences 3–72
physiology 91–95
Piazzi, Giuseppe 113
Pickering, Edward Charles 118
pineal body 94

Planck, Max 39, 105
planets 63–64
plant cells 95–96, 115
Plante, Gaston 116
plastics 31
platinum 112
Plato 5
Plücker, Julius 118
plutonism (geology) 69
plutonium 112, 119
polarization 121
polarized light 29, 57
polonium 112
potash 16–18
potassium 18, 111
power stations 53
praseodymium 112
preformation 78
Priestley, Joseph xiv, 6
Principles of Geology (Charles Lyell)
  72, 76, 78, 115
prisms 26
Procyon (star) 62
promethium 112
protactinium 112
Proust, Joseph Louis 7
Prout, William 114
pterodactyls 113
punched cards 113
Purkinje, Jan Evangelista 95

Q

quantitative analysis 6
quinine 30, 92

R

rabies 103
radioactivity 119
radiometers 118
radio waves 118
radium 112, 119
radon 112

Ramsay, William 28, 118
Ray, John 78
Rayleigh, Baron (Robert John
  Strutt) 28
recessive genes 117
reflector telescopes 63
reflex action 94
refraction 57, 121
Rensselaer Polytechnic Institute
  114
research laboratories 53
Rhazes 5
rhenium 112
rhodium 112
Ritter, Johann 55, 113
Roentgen, Wilhelm Konrad 119
Romantic movement xvi, xvii,
  70–71
Rosetta Stone 56
Rosse, Lord (William Parsons)
  63, 116
rotation of the Earth 116
Royal Astronomical Society (Great
  Britain) 114
Royal Institution (Great Britain)
  15–16
rubber gloves 101
rubidium 27, 112
Rumford, Count (Benjamin
  Thompson) 16, 36, 56
ruthenium 112
Rutherford, Ernest 51

S

safety elevators 116
safety lamps 114
samarium 112
Saturn (planet) 54, 63
scandium 28, 112
Schleiden, Matthias Jakob 91,
  95–96, 115
Schliemann, Heinrich 117

Schönbein, Christian 30
Schwabe, Samuel Heinrich 69,
  115
Schwann, Theodor 91, 96, *96*,
  115
scientific method 107–110
scientific theory 110
Sedgwick, Adam 85
selenium 112
Semmelweiss, Ignaz Philipp
  100–101
sensory regions (brain) 94
Shaw, George Bernard 87
Sholes, Christopher 117
Siebold, Karl Theodor Ernst von
  96
silicon 21, 112
silk industry 99
silver 12, 14, 21, 111
Sirius (star) 62
61 Cygni (star) 60–61
smallpox 103
Smithsonian Institution 115
soda (chemistry) 18
sodium 18, 21, 65–66, 111
sodium hydroxide 18
solar eclipse 23
solitaire 25
Somerville, Mary 66–67
Somerville, William 66–67
Spallanzani, Lazzaro 99
species 78, 113 *See also* evolution
specific heat 114
spectral lines 65, 114, 116–118
spectroscope 26–28, 64–65, 67
spectrum 26–27
speed of light 117–118
sperm cells 96
spinal nerves 92
spontaneous generation 97–99,
  116
Staël, Anne Louise Germaine de
  xvi–xvii

stars 60–63, 113, 115, 118
steam engines 36, 38–39
steam locomotives 114
Steinmetz, Charles 52
Stephenson, George xiii, 114
sterile technique 100–101
Stickney, Angelina 62
storage batteries 116
strontium 18, 24, 64, 112
Strutt, Robert John *See* Rayleigh,
  Baron
strychnine 92
Sturgeon, William 49
sugar 93
sulfur 21, 24, 45, 112
Sun 65, 68–69, 116
sunspots 69, 115
surgery 101
survival of the fittest 80
swallowing 92

T

tantalum 112
Taoists 4
tartaric acid crystals 29, 99
technetium 112
telegraph 115–116
telephones 118
telescopes 60–63
tellurium 112
temperature 11, 37, 40, 42
terbium 112
thallium 112
theory, scientific 110
Theory of Heat (James Clerk
  Maxwell) 43
thermochemistry 115
thermodynamics 38–43
  and chemical change 118
  definition of 121
  first law of 38–39, 43, 115–116
  second law of 39–43, 116

Thompson, Benjamin *See* Rumford, Count
Thomson, J. J. 119
Thomson, William *See* Kelvin, Lord
thorium 112
three-color theory 56
thulium 112
tin 112
tinfoil 12, 14
titanium 112
tortoises 77
transformers 49
transverse waves 57, 114
Travers, Morris 28
Treatise on Electricity and Magnetism (James Clerk Maxwell) 55
Trevithick, Richard xiii, 113
triads 5, 24
tuberculosis 102
tungsten 112
Tyndall, John 51
typewriters 117

U

ultraviolet light 55, 113
uniformitarianism (geology) 72, 78, 115
uranium 112
Uranus (planet) 63
urea 29, 115

V

vaccination 103
valence 116
vanadium 112
Van't Hoff, Jacobus 32
vasomotor nerves 93
Venus (planet) 63
vinegar (acetic acid) 22, 29
Virchow, Rudolf Carl 88, *97*, 97–98, 116

virus 103, 121
"vital force" 29, 32
vitalism 99, 115, 117, 122
vivisection 91, 93
Volta, Alessandro xv, xvii, 12–15, 19, 46, 48, 113
voltaic cells 12, *13*, 15–16, 46
vomiting 92

W

Wallace, Alfred Russel xv–xvi, *83*, 84–85, 89–90, 116
water 5, 10, 22
Watt, James xiii, 35–36, 38
wavelength 27, 122
wave theory of light 56–57, 113–114
Wedgwood, Josiah 75
Wells, H. G. 87
Werner, Abraham Gottlob 69
Whewell, William xiii, 115
Wilberforce, Samuel 86–87
Willis, Thomas 94
wine industry 99
Wöhler, Friedrich 29, 115
Wolff, Caspar 78
Wollaston, William 49
Wordsworth, William 16

X

xenon 28, 112
X rays 119

Y

yeast 99
Yerkes Observatory 63
Young, Thomas 15, 36, 55–58, 113–114
Young-Helmholtz three-color theory 56
ytterbium 112
yttrium 112

Z

zinc 27, 65, 112

zirconium 112

Zoological Philosophy (Jean-Baptiste
de Monet de Lamarck) 113